Accelerating Server-Side Development with Fastify

A comprehensive guide to API development for building a
scalable backend for your web apps

Manuel Spigolon

Maksim Sinik

Matteo Collina

BIRMINGHAM—MUMBAI

Accelerating Server-Side Development with Fastify

Group Product Manager: Rohit Rajkumar

Publishing Product Manager: Pavan Ramchandani

Senior Editor: Hayden Edwards

Senior Content Development Editor: Rashi Dubey

Technical Editor: Simran Ali

Copy Editor: Safis Editing

Project Coordinator: Sonam Pandey

Proofreader: Safis Editing

Indexer: Pratik Shirodkar

Production Designer: Joshua Misquitta

Marketing Coordinator: Nivedita Pandey and Namita Velgekar

First published: June 2023
Production reference: 2090923

Published by Packt Publishing Ltd.
Livery Place
35 Livery Street
Birmingham
B3 2PB, UK.

ISBN 978-1-80056-358-2
www.packtpub.com

To all the people who tolerate the quirks of developers, especially mine.

– Manuel Spigolon

To my beloved family, your love and support have been a constant source of inspiration and motivation. To my dear wife, Sara, for being my partner, confidante, and biggest fan.

– Maksim Sinik

To my daughter, Zoe, may you one day read this book and find Fastify and Node.js ancient.

– Matteo Collina

The Story of Fastify

The most important thing to understand regarding Fastify is that it's a community first, then a framework. Let me tell you the story of how everything started.

Between 2015 and 2016, I led a team to deliver a high-performance project in London with NearForm (www.nearform.com): we optimized this project so much that we could make Node.js saturate the network cards of our **Amazon Elastic Compute Cloud (Amazon EC2) virtual machines (VMs)**. However, because all the available frameworks added too much overhead, we had to implement all the HTTP APIs using only Node.js core, making the code hard to maintain. I knew I had to create a new framework that added as little overhead as possible while retaining a good developer experience. I also knew this was a gargantuan effort that I could never implement alone. I decided that this new web framework would adopt an open governance model and that I would start developing it once I convinced another developer to join me in this effort.

In June 2016, I delivered a training course in Bologna on Node.js with Avanscoperta (avanscoperta. it). One of the attendees asked me a question, "*How do I get started in open source?*" I asked him if he wanted to help me build our framework for Node.js. His name is *Tomas Della Vedova* (https:// delvedor.dev/). In September, we landed the first commit of what would become Fastify: at the time, it was called "beo," short for "be open," to signify our commitment to open governance. The prototype strikes the similarity of the current API of Fastify:

```
const beo = require('.')()
const http = require('http')
const server = http.createServer(beo)


const schema = {
  out: {
    type: 'object',
    properties: {
      hello: {
        type: 'string'
      }
    }
  }
}


beo.get('/', schema, function (req, reply) {
  reply(null, { hello: 'world' })
})
```

```
beo.post('/', schema, function (req, reply) {
  reply(null, { hello: 'world' })
})


server.listen(8000, function (err) {
  if (err) {
    throw err
  }


  console.log(`server listening on ${server.address().port}`)
})
```

We knew we had to make our framework performant. We achieved this with three components:

1. `find-my-way` (`https://github.com/delvedor/find-my-way`): This is one of the most performant routers for Node.js; it's based on a radix prefix tree data structure. This module was written by Tomas.
2. `fast-json-stringify` (`https://github.com/fastify/fast-json-stringify`)
3. `` `pino` `` (`https://github.com/pinojs/pino`): This is the fastest logger for Node.js.

Around the same time, I worked on a microservices framework called Seneca (sponsored by NearForm at the time), trying to fix a critical bug in its plugin system. While that change did not land, I was able to extensively research the topic of plugins, how to load them, and what was required to create a framework. All this research resulted in the creation of Avvio (`https://github.com/fastify/avvio`), the Fastify plugin system.

The development of Fastify proceeded swiftly, and we quickly evolved our prototype. Fastify was adopted as its primary framework by one start-up, LetzDoIt, led by Luca Maraschi. Thanks to the support of NearForm and LetzDoIt, Tomas and I had many iterations.

One year later, at Node.js Interactive 2017, I presented *"Take Your HTTP Server to Ludicrous Speed"* (`https://www.youtube.com/watch?v=gltzZjKYK1I`). On March 6, 2018, after 39 releases, Tomas cut v1.0.0 of Fastify.

Now the Fastify team has 17 collaborators, and its community is flourishing. It's now part of the OpenJS Foundation as an at-large project. The community is based on a few fundamental principles:

- **Open governance**: The project is governed by its members and follows a "consensus-seeking" model.

- **If something could be a plugin, it should be**: The Fastify-rich ecosystem allows for everything.

- **We encourage active participation in the community**: No one will fix your bugs; as a result, we often ask whether the reporter of an issue would be willing to send a Pull Request.

I hope that by the end of this book, you'd like to join this incredible community and start contributing!

Matteo Collina

Platformatic co-founder and chief technology officer (CTO)

OpenJS Foundation CPC director

Fastify lead maintainer

Node.js Technical Steering Committee member

Contributors

About the authors

Manuel Spigolon is a senior backend developer at NearForm. He is one of the core maintainers on the Fastify team. Manuel has developed and maintained a complex API that serves more than 10 million users worldwide.

I would like to express my gratitude to my beloved Vanessa for her unwavering patience and support throughout this long and arduous writing journey. I must also mention the "Jasmins real verificato" for their invaluable assistance. Moreover, I extend my heartfelt appreciation to my parents, Vincenzo and Ginetta, for instilling in me the strength of will to conquer any obstacle. Last but not least, I am deeply grateful to the Fastify community, whose constant support and encouragement have enabled me to reach new heights and achieve many of my dreams.

Maksim Sinik is a senior engineering manager and a core maintainer of the Fastify framework. He has a decade of experience as a Node.js developer with a strong interest in backend scalability. He designed the architecture and led the development of several service-based **Software-as-a-Service** (**SaaS**) platforms across multiple industries that process hundreds of thousands of requests.

I want to take a moment to acknowledge the incredible support and encouragement my family has provided me throughout my writing journey. To my darling daughter, Dafne, thank you for being a source of inspiration and joy. Your innocent and pure outlook on life has reminded me of the beauty and magic in this world. Your presence has brought light and meaning to my life, and I am grateful for every moment we spend together. To my wife, Sara, your support and encouragement have been a constant source of strength and motivation. I could not have done this without you. This book reflects our family's love and dedication, and I am proud to share it with you. To my mother, Zdravka, your love and support have given me the courage and confidence to chase my dreams and pursue my passions.

Matteo Collina is the co-founder and CTO of `Platformatic.dev` who has the goal of removing all friction from backend development. He is also a prolific open source author in the JavaScript ecosystem, and the modules he maintains are downloaded more than 17 billion times a year. Previously, he was the chief software architect at NearForm, the best professional services company in the JavaScript ecosystem. In 2014, he defended his Ph.D. thesis titled *Application Platforms for the Internet of Things*. Matteo is a member of the Node.js Technical Steering Committee, focusing on streams, diagnostics, and HTTP. He is also the author of the fast logger, Pino, and the Fastify web framework. Matteo is a renowned international speaker after more than 60 conferences, including OpenJS World, Node.js Interactive, NodeConf.eu, NodeSummit, JSConf.Asia, WebRebels, and JsDay, to name just a few. Since August 2023, he also serves as a community director on the OpenJS Foundation. In the summer, he loves sailing the Sirocco.

Fastify would not exist today without my wife Anna, who sent me to English school 20 years ago, Cian O'Maidin and Richard Rodgers, who recruited me for NearForm in 2014, Alberto Brandolini, who created Avanscoperta, and Luca Maraschi, who believed in the framework from the first day and invested in its development.

About the reviewers

Kyriakos Markakis has worked as a software engineer and application architect. He acquired experience in developing web applications by working on a wide range of projects in e-government, online voting, digital education, media monitoring, digital advertising, travel, and banking industries for more than 12 years.

During that time, he had the opportunity to work with technologies such as Node.js, Vue.js, Svelte, Java, REST APIs, GraphQL, PostgreSQL, MongoDB, Redis, Asynchronous Messaging, Docker, and others. He has been working with Fastify for the past four years.

He contributes his knowledge by mentoring beginners on technologies through e-learning programs and by reviewing books.

He was a reviewer of *Node.js Design Patterns (Third Edition)*.

Ethan Arrowood is a software engineer and ski instructor from Summit County, Colorado, USA. He is an experienced open source contributor with notable work on Fastify, Undici, and Node.js core. Most recently, he's been working with WinterCG to enable web runtime interoperability.

Table of Contents

Part 1: Fastify Basics

1

2

3

Working with Routes 59

4

Exploring Hooks 93

5

Exploring Validation and Serialization   117

Part 2: Build a Real-World Project

6

Project Structure and Configuration Management   151

10

Deployment and Process Monitoring for a Healthy Application   241

11

Meaningful Application Logging   263

Part 3: Advanced Topics

12

From a Monolith to Microservices   287

13

Performance Assessment and Improvement   301

14

Developing a GraphQL API   327

15

Type-Safe Fastify 353

Preface

Fastify is a plugin-based web framework designed to be highly performant. It fosters code reuse, thereby improving your time to market.

This book is a complete guide to server-side app development in Fastify, written by the core contributors of Fastify. The book starts from the core concepts, continuing with a real-world project and even touching on more advanced topics to enable you to build highly maintainable and scalable backend applications. You'll learn how to develop real-world RESTful applications from design to deployment, and build highly reusable components.

By the end of this book, you'll be able to design, implement, deploy, and maintain an application written in Fastify, and you'll also have the confidence you need to develop plugins and APIs in Fastify to contribute back to the Fastify and open source communities.

Who this book is for

This book is for mid-to expert-level backend web developers who have already used other backend web frameworks and have worked with HTTP protocol and its peculiarities. Developers looking to migrate to Fastify and learn how it can be a good fit for their next project, avoid architecture pitfalls, and build highly responsive and maintainable API servers will also find this book useful. The book assumes knowledge of JavaScript programming, Node.js, and backend development.

What this book covers

Chapter 1, What Is Fastify?, introduces Fastify, its purpose, and how it can speed up the development process. It covers the basic syntax for starting an application, adding endpoints, and configuring the server.

Chapter 2, The Plugin System and the Boot Process, explains the importance of Fastify plugins for developers, highlighting their role in achieving reusability, code sharing, and proper encapsulation between Fastify instances. The chapter focuses on the declaration and implementation of a simple plugin, gradually adding more layers to explore their interaction and dependency management.

Chapter 3, Working with Routes, covers the importance of routes in applications and how to manage them effectively with Fastify. It explains how to handle URL parameters, body, and query strings and how the router works in Fastify. The chapter also highlights the significance of Fastify's support for async/await handlers and provides tips for avoiding pitfalls when using them.

Chapter 4, *Exploring Hooks*, explains how Fastify is different from other web frameworks because it uses hooks instead of middleware. Mastering these hooks is essential for building stable and production-ready server applications. The chapter provides an overview of the hooks and the life cycle of Fastify applications.

Chapter 5, *Exploring Validation and Serialization*, focuses on implementing secure endpoints with input validation and optimizing them with the serialization process in Fastify. The chapter emphasizes the importance of using Fastify's built-in tools to control and adapt the default setup to suit your application's logic. You will learn how to configure and use Fastify components to create straightforward APIs that clients can easily consume.

Chapter 6, *Project Structure and Configuration Management*, guides you through creating a real-world RESTful cloud-native application, putting into action what you've learned in previous chapters. You'll build a solid project structure that can be reused for future projects, utilizing community packages and creating custom plugins as needed.

Chapter 7, *Building a RESTful API*, focuses on building a real-world Fastify application by defining routes, connecting to data sources, implementing business logic, and securing endpoints. The chapter will cover essential parts of the application, including implementing routes, connecting to data sources, securing endpoints, and applying the **don't repeat yourself** (**DRY**) principle to make code more efficient.

Chapter 8, *Authentication and File Handling*, covers user authentication and file handling in our application. We'll begin by implementing a reusable JWT authentication plugin that will handle user authentication, sessions, and authorization. We'll also explore decorators to expose authenticated user data in route handlers. Next, we'll develop a plugin to handle file import/export in CSV format for user to-do tasks.

Chapter 9, *Application Testing*, focuses on testing in Fastify. It covers the integrated testing tools and how to use them. It also explains how to run tests in parallel without requiring an HTTP server.

Chapter 10, *Deployment and Process Monitoring for a Healthy Application*, explores different deployment options for our Fastify API, focusing on a monolith deployment using Docker, MongoDB, and Fly. io. We will also learn about Node.js metrics and how to monitor the health of our server using Prometheus and Grafana.

Chapter 11, *Meaningful Application Logging*, explains how to implement effective logging in your Fastify application to keep track of what's happening and monitor its performance. You'll discover how to set up a logging configuration that captures all the necessary information without logging sensitive data and how to interpret and analyze logs to identify and troubleshoot issues.

Chapter 12, *From a Monolith to Microservices*, discusses restructuring a monolithic application into multiple modules to minimize conflicts between different teams. The chapter discusses adding new routes without increasing project complexity, splitting the monolith into microservices, and using an API gateway to route relevant calls. The chapter also covers logging, monitoring, error handling, and addressing operator questions.

Chapter 13, Performance Assessment and Improvement, focuses on optimizing the performance of your Fastify application. You will learn how to measure your code's performance to avoid slow performance, identify bottlenecks in your code, and prevent production issues. You will be introduced to an instrumentation library to help you analyze how your server reacts to high-volume traffic. Additionally, you will learn how to interpret the measurements and take action to keep your server's performance healthy.

Chapter 14, Developing a GraphQL API, explains how to incorporate GraphQL into your Fastify application using a specialized plugin. As GraphQL becomes increasingly popular, more and more APIs are being built with this query language. By using Fastify's unique architecture and avoiding common mistakes, you can take full advantage of GraphQL in your application.

Chapter 15, Type-Safe Fastify, explores how Fastify's built-in TypeScript support can help developers write more robust and maintainable applications. With compile-time type safety and better code completion, type inference, and documentation, developers can catch errors early in the development process and avoid unexpected runtime errors, leading to more stable and reliable applications. Using TypeScript with Fastify can also provide an extra layer of protection to code, making deployments safer and giving developers more confidence when developing code.

To get the most out of this book

To fully grasp the concepts discussed in this book, it is assumed that you have a basic understanding of the HTTP protocol, including the different methods and status codes. Additionally, you are expected to be familiar with running a Node.js application and can install and manage Node.js packages using npm.

Software/hardware covered in the book	Operating system requirements
Fastify 4	Windows, macOS, or Linux

Additionally, since we will be using containerization extensively throughout the book, it is recommended to have Docker installed on your machine before starting.

If you are using the digital version of this book, we advise you to type the code yourself or access the code from the book's GitHub repository (a link is available in the next section). Doing so will help you avoid any potential errors related to the copying and pasting of code.

Download the example code files

You can download the example code files for this book from GitHub at `https://github.com/PacktPublishing/Accelerating-Server-Side-Development-with-Fastify`. If there's an update to the code, it will be updated in the GitHub repository.

We also have other code bundles from our rich catalog of books and videos available at `https://github.com/PacktPublishing/`. Check them out!

Download the color images

We also provide a PDF file that has color images of the screenshots and diagrams used in this book. You can download it here: `https://packt.link/df1Dm`

Conventions used

There are a number of text conventions used throughout this book.

`Code in text`: Indicates code words in text, database table names, folder names, filenames, file extensions, pathnames, dummy URLs, user input, and Twitter handles. Here is an example: "The `getSchemas()` method returns a JSON key-value pair where the key is the schema's `$id` and the value is the JSON schema itself."

A block of code is set as follows:

```
{
  "id": 1,
  "name": "Foo",
  "hobbies": [ "Soccer", "Scuba" ]
}
```

When we wish to draw your attention to a particular part of a code block, the relevant lines or items are set in bold:

```
"test": "tap --before=test/run-before.js test/**/**.test.js
--after=test/run-after.js --no-check-coverage",
"test:coverage": "tap --coverage-report=html --before=test/run-before.
js test/**/**.test.js --after=test/run-after.js",
```

Any command-line input or output is written as follows:

```
<absolute URI>#<local fragment>
```

Bold: Indicates a new term, an important word, or words that you see onscreen. For instance, words in menus or dialog boxes appear in **bold**. Here is an example: "It is crucial to keep in mind that you should take care of the response status code when you implement your error handler; otherwise, it will be **500 – Server Error** by default."

> Tips or Important notes
> Appear like this.

Get in touch

Feedback from our readers is always welcome.

General feedback: If you have questions about any aspect of this book, email us at `customercare@packtpub.com` and mention the book title in the subject of your message.

Errata: Although we have taken every care to ensure the accuracy of our content, mistakes do happen. If you have found a mistake in this book, we would be grateful if you would report this to us. Please visit www.`packtpub.com/support/errata` and fill in the form.

Piracy: If you come across any illegal copies of our works in any form on the internet, we would be grateful if you would provide us with the location address or website name. Please contact us at `copyright@packt.com` with a link to the material.

If you are interested in becoming an author: If there is a topic that you have expertise in and you are interested in either writing or contributing to a book, please visit `authors.packtpub.com`.

Download a free PDF copy of this book

Thanks for purchasing this book!

Do you like to read on the go but are unable to carry your print books everywhere?

Is your eBook purchase not compatible with the device of your choice?

Don't worry, now with every Packt book you get a DRM-free PDF version of that book at no cost.

Read anywhere, any place, on any device. Search, copy, and paste code from your favorite technical books directly into your application.

The perks don't stop there, you can get exclusive access to discounts, newsletters, and great free content in your inbox daily

Follow these simple steps to get the benefits:

1. Scan the QR code or visit the link below

https://packt.link/free-ebook/9781800563582

2. Submit your proof of purchase
3. That's it! We'll send your free PDF and other benefits to your email directly

Part 1:
Fastify Basics

In this part, you will learn about the framework's essential features and the philosophy behind it, and embrace "the Fastify way" mindset to speed up your team's development process, leveraging the unique plugin system it offers.

In this part, we cover the following chapters:

- *Chapter 1, What Is Fastify?*
- *Chapter 2, The Plugin System and the Boot Process*
- *Chapter 3, Working with Routes*
- *Chapter 4, Exploring Hooks*
- *Chapter 5, Exploring Validation and Serialization*

1
What Is Fastify?

Nowadays, building a solid application is just not enough, and the time it takes to get an application to market has become one of the major constraints a development team must deal with. For this reason, Node.js is the most used runtime environment currently adopted by companies. Node.js has proved how easy and flexible it is to build web applications. To compete in this tech scene that moves at the speed of light, you need to be supported by the right tools and frameworks to help you implement solid, secure, and fast applications. The pace is not only about the software performance, but it is also important to take the time to add new features and to keep the software reliable and extensible. Fastify gives you a handy development experience without sacrificing performance, security, and source readability. With this book, you will get all the knowledge to use this framework in the most profitable way.

This chapter will explain what Fastify is, why it was created, and how it can speed up the development process. You will become confident with the basic syntax to start your application, add your first endpoints, and learn how to configure your server to overview all the essential options.

You will start to explore all the features this framework offers, and you will get your hands dirty as soon as possible. There is a first basic example that we will implement to explain the peculiarities of the framework. We will analyze the environment configuration and how to shut down the application properly.

In this chapter, we will cover the following topics:

- What is Fastify?
- Starting your server
- Adding basic routes
- Adding a basic plugin instance
- Understanding configuration types
- Shutting down the application

Technical requirements

Before proceeding, you will need a development environment to write and execute your first Fastify code. You should configure:

- A text editor or an IDE such as VS Code

- Node.js v18 or above (you can find this here: `https://nodejs.org/it/download/`)

- An HTTP client to test out code; you may use a browser, CURL, or Postman

All the code examples in this chapter may be found on GitHub at `https://github.com/PacktPublishing/Accelerating-Server-Side-Development-with-Fastify/tree/main/Chapter%201`.

What is Fastify?

Fastify is a Node.js web framework used to build server applications. It helps you develop an HTTP server and create your API in an easy, fast, scalable, and secure way!

It was born in late 2016, and since its first release in 2018, it has grown at the speed of light. It joined the *OpenJS Foundation* as an At-Large project in 2020 when it reached version 3, which is the version we are going to work with!

This framework focuses on unique aspects that are uncommon in most other web frameworks:

- Improvement of the developer experience: This streamlines their work and promotes a **plugin design system**. This architecture helps you structure your application in smaller pieces of code and apply good programming patterns such as **DRY (Don't Repeat Yourself)**, **Immutability**, and **Divide & Conquer**.

- Comprehensive performance: This framework is built to be the fastest.

- Up to date with the evolution of the Node.js runtime: This includes quick bugfixes and feature delivery.

- Ready to use: Fastify helps you set up the most common issues you may face during the implementation, such as application logging, security concerns, automatic test implementation, and user input parsing.

- Community-driven: Supports and listens to the framework users.

The result is a flexible and highly extensible framework that will lead you to create reusable components. These concepts give you the boost to develop a **proof of concept** (**PoC**) or large applications faster and faster. Creating your plugin system takes less time to meet the business need without losing the possibility to create an excellent code base and a performant application.

Moreover, Fastify has a clear **Long Term Support** (**LTS**) policy that supports you while planning the updates of your platform and that stays up to date with the Node.js versions and features.

Fastify provides all these aspects to you through a small set of components that we are going to look at in the next section.

Fastify's components

Fastify makes it easier to create and manage an HTTP server and the HTTP request lifecycle, hiding the complexity of the Node.js standard modules. It has two types of components: the **main components** and **utility elements**. The former comprise the framework, and it is mandatory to deal with them to create an application. The latter includes all the features you may use at your convenience to improve the code reusability.

The main components that are going to be the main focus of this book and that we are going to discuss in this chapter are:

- The **root application instance** represents the Fastify API at your disposal. It manages and controls the standard Node.js `http.Server` class and sets all the endpoints and the default behavior for every request and response.

- A **plugin instance** is a child object of the application instance, which shares the same interface. It isolates itself from other sibling plugins to let you build independent components that can't modify other contexts. *Chapter 2* explores this component in depth, but we will see some examples here too.

- The `Request` object is a wrapper of the standard Node.js `http.IncomingMessage` that is created for every client's call. It eases access to the user input and adds functional capabilities, such as logging and client metadata.

- The `Reply` object is a wrapper of the standard Node.js `http.ServerResponse` and facilitates sending a response back to the user.

The utility components, which will be further discussed in *Chapter 4* are:

- The **hook** functions that act, when needed, during the lifecycle of the application or a single request and response

- The **decorators**, which let you augment the features installed by default on the main components, avoiding code duplication

- The **parsers**, which are responsible for the request's payload conversion to a primitive type

That's all! All these parts work together to provide you with a toolkit that supports you during every step of your application lifetime, from prototyping to testing, without forgetting the evolution of your code base to a manageable one.

> **Many names for one component**
>
> It is essential to learn the component's name, especially the plugin instance one. It has many synonyms, and the most common are plugin, instance, or child instance. The Fastify official documentation uses these terms broadly and interchangeably, so it is beneficial to keep them all in mind.

We have read about all the actors that build Fastify's framework and implement its focus aspects. Thanks to this quick introduction, you know their names, and we will use them in the following sections. The following chapters will further discuss every component and unveil their secrets.

Starting your server

Before we start with Fastify, it is necessary to set up a developing environment. To create an empty project with npm, open your system's shell and run the following commands:

```
mkdir fastify-playground
cd fastify-playground/
npm init --yes
npm install fastify
```

These commands create an empty folder and initialize a Node.js project in the new directory; you should see a successful message on each npm <command> execution.

Now, we are ready to start an HTTP server with Fastify, so create a new starts.cjs file and check out these few lines:

```
const fastify = require('fastify') // [1]
const serverOptions = { // [2]
  logger: true
}
const app = fastify(serverOptions) // [3]
app.listen({
  port: 8080,
  host: '0.0.0.0'
})
  .then((address) => { // [4]
    // Server is now listening on ${address}
  })
```

Let's break up each of the elements of this code. The imported framework is a factory function [1] that builds the Fastify server **root application instance**.

Book code style

All the book's code snippets are written in **CommonJS (CJS)**. The CJS syntax has been preferred over **ECMAScript Modules (ESM)** because it is not yet fully supported by tools such as **application performance monitoring (APM)** or test frameworks. Using the `require` function to import the modules lets us focus on code, avoiding issues that can't be covered in this book.

The factory accepts an optional JavaScript object input [2] to customize the server and its behavior—for instance, supporting HTTPS and the HTTP2 protocol. You will get a complete overview of this matter later on in this chapter. The application instance, returned by the factory, lets us build the application by adding routes to it, configuring and managing the HTTP server's start and stop phases.

After the server has built our instance [3], we can execute the `listen` method, which will return a `Promise`. Awaiting it will start the server [4]. This method exposes a broad set of interfaces to configure where to listen for incoming requests, and the most common is to configure the PORT and HOST.

---

listen

Calling `listen` with the `0.0.0.0` host will make your server accept any unspecified IPv4 addresses. This configuration is necessary for a Docker container application or in any application that is directly exposed on the internet; otherwise, external clients won't be able to call your HTTP server.

To execute the previous code, you need to run this command:

```
node starts.cjs
```

This will start the Fastify server, and calling the `http://localhost:8080/` URL with an HTTP client or just a browser must show a 404 response because we didn't add any route yet.

Congratulations—you have started your first Fastify server! You can kill it by pressing the *Ctrl + C* or *Cmd+ C* buttons.

We have seen the root instance component in action. In a few lines of code, we were able to start an HTTP server with no burden! Before continuing to dig into the code, in the next section, we will start understanding what Fastify does under the hood when we start it.

Lifecycle and hooks overview

Fastify implements two systems that regulate its internal workflow: the **application lifecycle** and the **request lifecycle**. These two lifecycles trigger a large set of events during the application's lifetime. Listening to those events will let us customize the data flow around the endpoints or simply add monitoring tools.

The application lifecycle tracks the status of the application instance and triggers this set of events:

- The `onRoute` event acts when you add an endpoint to the server instance

- The `onRegister` event is unique as it performs when a new **encapsulated context** is created

- The `onReady` event runs when the application is ready to start listening for incoming HTTP requests

- The `onClose` event executes when the server is stopping

All these events are Fastify's hooks. More specifically, a function that runs whenever a specific event happens in the system is a **hook**. The hooks that listen for application lifecycle events are called **application hooks**. They can intercept and control the application server boot phases, which involve:

- The routes' and plugins' initialization

- The application's start and close

Here is a quick usage example of what happens after adding this code before the `listen` call in the previous code block:

```
app.addHook('onRoute', function inspector(routeOptions) {
  console.log(routeOptions)
})
app.addHook('onRegister', function inspector(plugin,
pluginOptions) {
  console.log('Chapter 2, Plugin System and Boot Process')
})
app.addHook('onReady', function preLoading(done) {
  console.log('onReady')
  done()
})
app.addHook('onClose', function manageClose(done) {
  console.log('onClose')
  done()
})
```

We see that there are two primary API interfaces for these hooks:

- The onRoute and the onRegister hooks have some object arguments. These types can only manipulate the input object adding side effects. A side effect changes the object's properties value, causing new behavior of the object itself.

- The onReady and onClose hooks have a callback style function input instead. The done input function can impact the application's startup because the server will wait for its completion until you call it. In this timeframe, it is possible to load some external data and store it in a local cache. If you call the callback with an error object as the first parameter, done(new Error()), the application will listen, and the error will bubble up, crashing the server startup. So, it's crucial to load relevant data and manage errors to prevent them from blocking the server.

As presented in the preceding example, running our source code will print out only the onReady string in the console. Why are our hooks not running? This happens because the events we are listening to are not yet triggered. They will start working by the end of this chapter!

Note that whenever a Fastify interface exposes a done or next argument, you can omit it and provide an async function instead. So, you can write:

```
app.addHook('onReady', async function preLoading() {
  console.log('async onReady')
  // the done argument is gone!
})
```

If you don't need to run async code execution such as I/O to the filesystem or to an external resource such as a database, you may prefer the callback style. It provides a simple function done within the arguments, and is slightly more performant than an async function!

You can call the addHook() method multiple times to queue the hooks' functions. Fastify guarantees to execute them all in the order of addition.

All these phases can be schematized into this execution flow:

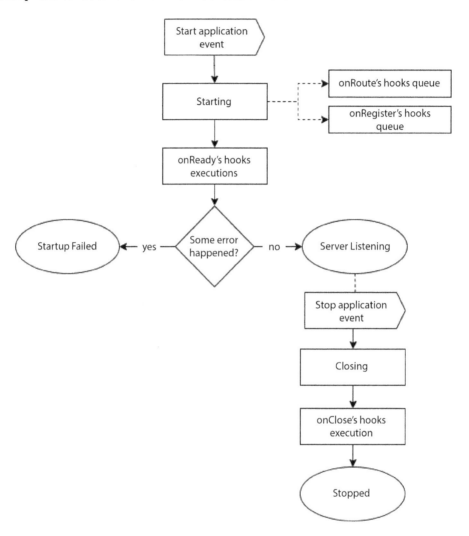

Figure 1.1 – Application lifecycle

At the start of the application, the onRoute and onRegister hooks are executed whenever a new route or a new encapsulated context is created (we will discuss the encapsulated context by the end of this chapter, in the *Adding a basic plugin instance* section). The dashed lines in *Figure 1.1* mean that these hooks' functions are run synchronously and are not awaited before the server starts up. When the application is loaded, the onReady hooks queue is performed, and the server will start listening if there are no errors during this startup phase. Only after the application is up and running will it be able to receive stop events. These events will start the closing stage, during which the onClose hooks' queue will be executed before stopping the server. The closing phase will be discussed in the *Shutting down the application* section.

The request lifecycle, instead, has a lot more events. But keep calm—*Chapter 4* talks about them extensively, and you will learn how to use them, why they exist, and when you should use them. The hooks listening to the request's lifecycle events are **request and reply hooks**. This lifecycle defines the flow of every HTTP request that your server will receive. The server will process the request in two phases:

- The routing: This step must find the function that must evaluate the request
- The handling of the request performs a set of events that compose the request lifecycle

The request triggers these events in order during its handling:

1. `onRequest`: The server receives an HTTP request and routes it to a valid endpoint. Now, the request is ready to be processed.
2. `preParsing` happens before the evaluation of the request's body payload.
3. The `preValidation` hook runs before applying **JSON Schema validation** to the request's parts. Schema validation is an essential step of every route because it protects you from a malicious request payload that aims to leak your system data or attack your server. *Chapter 5* discusses this core aspect further and will show some harmful attacks.
4. `preHandler` executes before the endpoint handler.
5. `preSerialization` takes action before the response payload transformation to a String, a Buffer, or a Stream, in order to be sent to the client.
6. `onError` is executed only if an error happens during the request lifecycle.
7. `onSend` is the last chance to manipulate the response payload before sending it to the client.
8. `onResponse` runs after the HTTP request has been served.

We will see some examples later on. I hope you have enjoyed the spoilers! But first, we must deep dive into the Fastify server to understand how to use it and how it interacts with the lifecycle.

The root application instance

The root application instance is the main API you need to create your API. All the functions controlling the incoming client's request must be registered to it, and this provides a set of helpers that let you best organize the application. We have already seen how to build it using the `const app = fastify(serverOptions)` statement. Now, we will present a general overview of the possible options to configure and use this object.

Server options

When you create a Fastify server, you have to choose some key aspects before starting the HTTP server. You may configure them, providing the option input object, which has many parameters listed in the Fastify documentation (`https://www.fastify.io/docs/latest/Reference/Server/`).

Now, we will explore all the aspects you can set with this configuration:

- The `logger` parameter gives you the control to adapt the default logger to your convenience and system infrastructure to archive distributed logging and meaningful logs; *Chapter 11* will discuss broadly how to best set up these parameters.

- `https: object` sets up the server to listen for **Transport Layer Security** (**TLS**) sockets. We will see some examples later on in *Chapter 7*.

- `keepAliveTimeout`, `connectionTimeout`, and `http2SessionTimeout` are several timeout parameters after which the HTTP request socket will be destroyed, releasing the server resources. These parameters are forwarded to the standard Node.js `http.Server`.

- Routing customization to provide stricter or laxer constraints—for instance, a case-insensitive URL and more granular control to route a request to a handler based on additional information, such as a request header instead of an HTTP method and HTTP URL. We will cover this in *Chapter 3*.

- `maxParamLength: number<length>` limits the path parameter string length.

- `bodyLimit: number<byte>` caps the request body payload size.

- `http2: boolean` starts an HTTP2 server, which is useful to create a long-lived connection that optimizes the exchange of data between client and server.

- The `ajv` parameter tweaks the validation defaults to improve the fit of your setup. *Chapter 5* will show you how to use it.

- The `serverFactory: function` manages the low-level HTTP server that is created. This feature is a blessing when working in a serverless environment.

- The `onProtoPoisoning` and `onConstructorPoisoning` default security settings are the most conservative and provide you with an application that's secure by default. Changing them is risky and you should consider all the security issues because it impacts the default request body parser and can lead to code injection. *Chapter 4* will show you an example of these parameters in action.

Are you overwhelmed by all these options? Don't worry. We are going to explore some of them with the following examples. The options provided not only allow you to adapt Fastify to a wide range of general use cases but extend this possibility to edge cases as well; usually, you may not need to configure all these parameters at all. Just remember that default settings are ready for production and provide the most secure defaults and the most useful utilities, such as `404 Not Found` and `500 Error` handlers.

Application instance properties

The Fastify server exposes a set of valuable properties to access:

- An `app.server` getter that returns the Node.js standard `http.Server` or `https.Server`.

- `app.log` returns the application logger that you can use to print out meaningful information.

- `app.initialConfig` to access the input configuration in read-only mode. It will be convenient for plugins that need to read the server configuration.

We can see them all in action at the server startup:

```
await app.listen({
  port: 0,
  host: '0.0.0.0'
})
app.log.debug(app.initialConfig, 'Fastify listening with
the config')
const { port } = app.server.address()
app.log.info('HTTP Server port is %i', port)
```

Setting the port parameter to 0 will ask the operating system to assign an unused host's port to your HTTP server that you can access through the standard Node.js `address()` method. Running the code will show you the output log in the console, which shows the server properties.

Unfortunately, we won't be able to see the output of the `debug` log. The log doesn't appear because Fastify is protecting us from misconfiguration, so, by default, the log level is at `info`. The log-level values accepted by default are `fatal`, `error`, `warn`, `info`, `debug`, `trace`, and `silent`. We will see a complete log setup in *Chapter 11*.

So, to fix this issue, we just need to update our `serverConfig` parameter to the following:

```
const serverOptions = {
  logger: {
    level: 'debug'
  }
}
```

By doing so, we will see our log printed out on the next server restart! We have seen the instance properties so far; in the next section, we will introduce the server instance methods.

Application instance methods

The application instance lets us build the application, adding routes and empowering Fastify's default components. We have already seen the `app.addHook(eventName, hookHandler)` method, which appends a new function that runs whenever the **request lifecycle** or the **application lifecycle** triggers the registered event.

The methods at your disposal to create your application are:

- `app.route(options[, handler])` adds a new endpoint to the server.
- `app.register(plugin)` adds plugins to the server instance, creating a new server context if needed. This method provides Fastify with **encapsulation**, which will be covered in *Chapter 2*.
- `app.ready([callback])` loads all the applications without listening for the HTTP request.
- `app.listen(port|options [,host, callback])` starts the server and loads the application.
- `app.close([callback])` turns off the server and starts the closing flow. This generates the possibility to close all the pending connections to a database or to complete running tasks.
- `app.inject(options[, callback])` loads the server until it reaches the ready status and submits a mock HTTP request. You will learn about this method in *Chapter 9*.

This API family will return a native `Promise` if you don't provide a callback parameter. This code pattern works for every feature that Fastify provides: whenever there is a callback argument, you can omit it and get back a promise instead!

Now, you have a complete overview of the Fastify server instance component and the lifecycle logic that it implements. We are ready to use what we have read till now and add our first endpoints to the application.

Adding basic routes

The routes are the entry to our business logic. The HTTP server exists only to manage and expose routes to clients. A route is commonly identified by the HTTP method and the URL. This tuple matches your function handler implementation. When a client hits the route with an HTTP request, the function handler is executed.

We are ready to add the first routes to our playground application. Before the `listen` call, we can write the following:

```
app.route({
  url: '/hello',
  method: 'GET',
  handler: function myHandler(request, reply) {
    reply.send('world')
```

```
  }
})
```

The route method accepts a JavaScript object as a parameter to set the HTTP request handler and the endpoint coordinates. This code will add a GET /hello endpoint that will run the myHandler function whenever an HTTP request matches the HTTP method and the URL that was just set. The handler should implement the business logic of your endpoint, reading from the request component and returning a response to the client via the reply object.

Note that running the previous code in your source code must trigger the onRoute hook that was sleeping before; now, the http://localhost:8080/hello URL should reply, and we finally have our first endpoint!

Does the onRoute hook not work?

If the onRoute hook doesn't show anything on the terminal console, remember that the addRoute method must be called after the addHook function! You have spotted the nature a hook may have: the application's hooks are synchronous and are triggered as an event happens, so the order of the code matters for these kinds of hooks. This topic will be broadly discussed in *Chapter 4*.

When a request comes into the Fastify server, the framework takes care of the routing. It acts by default, processing the HTTP method and the URL from the client, and it searches for the correct handler to execute. When the router finds a matching endpoint, the request lifecycle will start running. Should there be no match, the default 404 handler will process the request.

You have seen how smooth adding new routes is, but can it be even smoother? Yes, it can!

Shorthand declaration

The HTTP method, the URL, and the handler are mandatory parameters to define new endpoints. To give you a less verbose routes declaration, Fastify supports three different shorthand syntaxes:

```
app.get(url, handlerFunction) // [1]
app.get(url, { // [2]
  handler: handlerFunction,
  // other options
})
app.get(url, [options], handlerFunction) // [3]
```

The first shorthand [1] is the most minimal because it accepts an input string as a URL and handler. The second shorthand syntax [2] with options will expect a string URL and a JavaScript object as input with a `handler` key with a function value. The last one [3] mixes the previous two syntaxes and lets you provide the string URL, route options, and function handler separately: this will be useful for those routes that share the same options but not the same handler!

All the HTTP methods, including GET, POST, PUT, HEAD, DELETE, OPTIONS, and PATCH, support this declaration. You need to call the correlated function accordingly: `app.post()`, `app.put()`, `app.head()`, and so on.

The handler

The route handler is the function that must implement the endpoint business logic. Fastify will provide your handlers with all its main components, in order to serve the client's request. The `request` and `reply` object components will be provided as arguments, and provide the server instance through the function binding:

```
function business(request, reply) {
  // `this` is the Fastify application instance
  reply.send({ helloFrom: this.server.address() })
}
app.get('/server', business)
```

Using an arrow function will prevent you from getting the function context. Without the context, you don't have the possibility to use the `this` keyword to access the application instance. The arrow function syntax may not be a good choice because it can cause you to lose a great non-functional feature: the source code organization! The following handler will throw a `Cannot read property 'server' of undefined` error:

```
app.get('/fail', (request, reply) => {
  // `this` is undefined
  reply.send({ helloFrom: this.server.address() })
})
```

> **Context tip**
>
> It would be best to choose named functions. In fact, avoiding arrow function handlers will help you debug your application and split the code into smaller files without carrying boring stuff, such as the application instance and logging objects. This will let you write shorter code and make it faster to implement new endpoints. The context binding doesn't work exclusively on handlers but also works on every Fastify input function and hook, for example!

The business logic can be synchronous or asynchronous: Fastify supports both interfaces, but you must be aware of how to manage the `reply` object in your source code. In both situations, the handler should never call `reply.send(payload)` more than once. If this happens, it will work just for the first call, while the subsequent call will be ignored without blocking the code execution:

```
app.get('/multi', function multi(request, reply) {
  reply.send('one')
  reply.send('two')
  reply.send('three')
  this.log.info('this line is executed')
})
```

The preceding handler will reply with the `one` string, and the next `reply.send` calls will log an `FST_ERR_REP_ALREADY_SENT` error in the console.

To ease this task, Fastify supports the return even in the synchronous function handler. So, we will be able to rewrite our first section example as the following:

```
function business(request, reply) {
  return { helloFrom: this.server.address() }
}
```

Thanks to this supported interface, you will not mess up multiple `reply.send` calls!

The async handler function may completely avoid calling the `reply.send` method instead. It can return the payload directly. We can update the GET `/hello` endpoint to this:

```
app.get('/hello', async function myHandler(request, reply) {
  return 'hello' // simple returns of a payload
})
```

This change will not modify the output of the original endpoint: we have updated a synchronous interface to an async interface, updating how we manage the response payload accordingly. The async functions that do not execute the `send` method can be beneficial to reuse handlers in other handler functions, as in the following example:

```
async function foo (request, reply) {
  return { one: 1 }
}
async function bar (request, reply) {
  const oneResponse = await foo(request, reply)
  return {
    one: oneResponse,
    two: 2
  }
}
```

```
app.get('/foo', foo)
app.get('/bar', bar)
```

As you can see, we have defined two named functions: `foo` and `bar`. The `bar` handler executes the `foo` function and it uses the returned object to create a new response payload.

Avoiding the `reply` object and returning the response payload unlocks new possibilities to reuse your handler functions, because calling the `reply.send()` method would explicitly prevent manipulating the results as the `bar` handler does.

Note that a sync function may return a `Promise` chain. In this case, Fastify will manage it like an async function! Look at this handler, which will return file content:

```
const fs = require('fs/promises')
app.get('/file', function promiseHandler(request, reply) {
  const fileName = './package.json'
  const readPromise = fs.readFile(fileName, { encoding:
  'utf8' })
  return readPromise
})
```

In this example, the handler is a sync function that returns `readPromise:Promise`. Fastify will wait for its execution and reply to the HTTP request with the payload returned by the promise chain. Choosing the async function syntax or the `sync` and `Promise` one depends on the output. If the content returned by the `Promise` is what you need, you can avoid adding an extra async function wrapper, because that will slow down your handler execution.

The Reply component

We have already met the `Reply` object component. It forwards the response to the client, and it exposes all you need in order to provide a complete answer to the request. It provides a full set of functions to control all response aspects:

- `reply.send(payload)` will send the response payload to the client. The payload can be a String, a JSON object, a Buffer, a Stream, or an Error object. It can be replaced by returning the response's body in the handler's function.
- `reply.code(number)` will set the response status code.
- `reply.header(key, value)` will add a response header.
- `reply.type(string)` is a shorthand to define the Content-Type header.

The `Reply` component's methods can be chained to a single statement to reduce the code noise as follows: `reply.code(201).send('done')`.

Another utility of the `Reply` component is the headers' auto-sense. `Content-Length` is equal to the length of the output payload unless you set it manually. `Content-Type` resolves strings to `text/plain`, a JSON object to `application/json`, and a stream or a buffer to the `application/octet-stream` value. Furthermore, the HTTP return status is 200 Successful when the request is completed, whereas when an error is thrown, 500 Internal Server Error will be set.

If you send a `Class` object, Fastify will try to call `payload.toJSON()` to create an output payload:

```
class Car {
  constructor(model) {
    this.model = model
  }
  toJSON() {
    return {
      type: 'car',
      model: this.model
    }
  }
}
app.get('/car', function (request, reply) {
  return new Car('Ferrari')
})
```

Sending a response back with a new `Car` instance to the client would result in the JSON output returned by the `toJSON` function implemented by the class itself. This is useful to know if you use patterns such as **Model View Controller (MVC)** or **Object Relational Mapping (ORM)** extensively.

The first POST route

So far, we have seen only HTTP GET examples to retrieve data from the backend. To submit data from the client to the server, we must switch to the POST HTTP method. Fastify helps us read the client's input because the JSON input and output is a first-class citizen, and to process it, we only need to access the `Request` component received as the handler's argument:

```
const cats = []
app.post('/cat', function saveCat(request, reply) {
  cats.push(request.body)
  reply.code(201).send({ allCats: cats })
})
```

This code will store the request body payload in an in-memory array and send it back as a result.

Calling the `POST /cat` endpoint with your HTTP client will be enough to parse the request's payload and reply with a valid JSON response! Here is a simple request example made with `curl`:

```
$ curl --request POST "http://127.0.0.1:8080/cat" --header "Content-
Type: application/json" --data-raw "{\"name\":\"Fluffy\"}"
```

The command will submit the `Fluffy` cat to our endpoint, which will parse the payload and store it in the `cats` array.

To accomplish this task, you just have to access the `Request` component without dealing with any complex configuration or external module installation! Now, let's explore in depth the `Request` object and what it offers out of the box.

The Request component

During the implementation of the POST route, we read the `request.body` property. The body is one of the most used keys to access the HTTP request data. You have access to the other piece of the request through the API:

- `request.query` returns a key-value JavaScript object with all the query-string input parameters.
- `request.params` maps the URL path parameters to a JavaScript object.
- `request.headers` maps the request's headers to a JavaScript object as well.
- `request.body` returns the request's body payload. It will be a JavaScript object if the request's Content-Type header is `application/json`. If its value is `text/plain`, the body value will be a string. In other cases, you will need to create a parser to read the request payload accordingly.

The `Request` component is capable of reading information about the client and the routing process too:

```
app.get('/xray', function xRay(request, reply) {
  // send back all the request properties
  return {
    id: request.id, // id assigned to the request in req-
                    <progress>
    ip: request.ip, // the client ip address
    ips: request.ips, // proxy ip addressed
    hostname: request.hostname, // the client hostname
    protocol: request.protocol, // the request protocol
    method: request.method, // the request HTTP method
    url: request.url, // the request URL
    routerPath: request.routerPath, // the generic handler
                                    URL
    is404: request.is404 // the request has been routed or
                          not
```

```
    }
})
```

`request.id` is a string identifier with the `"req-<progression number>"` format that Fastify assigns to each request. The progression number restarts from 1 at every server restart. The ID's purpose is to connect all the logs that belong to a request:

```
app.get('/log', function log(request, reply) {
    request.log.info('hello') // [1]
    request.log.info('world')
    reply.log.info('late to the party') // same as
                                         request.log

    app.log.info('unrelated') // [2]
    reply.send()
})
```

Making a request to the GET `/log` endpoint will print out to the console six logs:

- Two logs from Fastify's default configuration that will trace the incoming request and define the response time

- Four logs previously written in the handler

The output should be as follows:

```
{"level":30,"time":1621781167970,"pid":7148,"hostname":"
EOMM-XPS","reqId":"req-1","req":{"method":"GET","url":"/
log","hostname":"localhost:8080","remoteAddress":"127.
0.0.1","remotePort":63761},"msg":"incoming request"}
{"level":30,"time":1621781167976,"pid":7148,"hostname":"EOMM-
XPS","reqId":"req-1","msg":"hello"}
{"level":30,"time":1621781167977,"pid":7148,"hostname":"EOMM-
XPS","reqId":"req-1","msg":"world"}
{"level":30,"time":1621781167978,"pid":7148,"hostname":"EOMM-
XPS","reqId":"req-1","msg":"late to the party"}
{"level":30,"time":1621781167979,"pid":7148,"hostname":"EOMM-
XPS","msg":"unrelated"}
{"level":30,"time":1621781167991,"pid":7148,"hostname":"EOMM-
XPS","reqId":"req-1","res":{"statusCode":200},"responseTime":17.831200
003623962,"msg":"request completed"}
```

Please note that only the `request.log` and `reply.log` commands [1] have the `reqId` field, while the application logger doesn't [2].

The request ID feature can be customized via these server options if it doesn't fit your system environment:

```
const app = fastify({
  logger: true,
  disableRequestLogging: true, // [1]
  requestIdLogLabel: 'reqId', // [2]
  requestIdHeader: 'request-id', // [3]
  genReqId: function (httpIncomingMessage) { // [4]
  return `foo-${Math.random()}`
  }
})
```

By turning off the request and response logging [1], you will take ownership of tracing the clients' calls. The [2] parameter customizes the field name printed out in the logs, and [3] informs Fastify to obtain the ID to be assigned to the incoming request from a specific HTTP header. When the header doesn't provide an ID, the `genReqId` function [4] must generate a new ID.

The default log output format is a JSON string designed to be consumed by external software to let you analyze the data. This is not true in a development environment, so to see a human-readable output, you need to install a new module in the project:

```
npm install pino-pretty --save-dev
```

Then, update your logger settings, like so:

```
const serverOptions = {
  logger: {
    level: 'debug',
    transport: {
      target: 'pino-pretty'
    } }
}
```

Restarting the server with this new configuration will instantly show a nicer output to read. The logger configuration is provided by `pino`. Pino is an external module that provides the default logging feature to Fastify. We will explore this module too in *Chapter 11*.

Parametric routes

To set a path parameter, we must write a special URL syntax, using the colon before our parameter's name. Let's add a GET endpoint beside our previous POST /cat route:

```
app.get('/cat/:catName', function readCat(request, reply) {
  const lookingFor = request.params.catName
  const result = cats.find(cat => cat.name == lookingFor)
  if (result) {
```

```
    return { cat: result }
  } else {
    reply.code(404)
    throw new Error(`cat ${lookingFor} not found`)
  }
})
```

This syntax supports regular expressions too. For example, if you want to modify the route previously created to exclusively accept a numeric parameter, you have to write the RegExp string at the end of the parameter's name between parentheses:

```
app.get('/cat/:catIndex(\\d+)', function readCat(request,
reply) {
  const lookingFor = request.params.catIndex
  const result = cats[lookingFor]
  // …
})
```

Adding the regular expression to the parameter name will force the router to evaluate it to find the right route match. In this case, only when `catIndex` is a number will the handler be executed; otherwise, the 404 fallback will take care of the request.

> **Regular expression pitfall**
>
> Don't abuse the regular expression syntax in the path parameters because it comes with a performance cost. Moreover, a mismatch of regular expressions will lead to a 404 response. You may find it useful to validate the parameter with the Fastify validator, which we present in *Chapter 5* to reply with a `400 Bad Request` status code.

The Fastify router supports the wildcard syntax too. It can be useful to redirect a root path or to reply to a set of routes with the same handler:

```
app.get('/cat/*', function sendCats(request, reply) {
  reply.send({ allCats: cats })
})
```

Note that this endpoint will not conflict with the previous because they are not overlapping, thanks to the match order:

1. Perfect match: `/cat`

2. Path parameter match: `/cat/:catIndex`

3. Wildcards: `/cat/*`

4. Path parameter with a regular expression: `/cat/:catIndex(\\d+)`

Under the hood, Fastify uses the find-my-way package to route the HTTP request, and you can benefit from its features.

This section explored how to add new routes to our application and how many utilities Fastify gives us, from application logging to user input parsing. Moreover, we covered the high flexibility of the reply object and how it supports us when returning complex JSON to the client. We are now ready to go further and start understanding Fastify plugin system basics.

Adding a basic plugin instance

Previously, we talked about a plugin instance as a child component of an application instance.

To create one, you simply need to write the following:

```
app.register(function myPlugin(pluginInstance, opts, next) {
  pluginInstance.log.info('I am a plugin instance,
    children of app')
  next()
}, { hello: 'the opts object' })
```

These simple lines have just created an **encapsulated context**: this means that every event, hook, plugin, and decorator registered in the myPlugin function scope will remain inside that context and all its children. Optionally, you can provide an input object as a second parameter to the register function. This will propagate the input to the plugin's opts parameter. If you move the plugin to another file, this will become extremely useful when sharing a configuration through files.

To see how the encapsulated context acts, we can investigate the output of the following example:

```
app.addHook('onRoute', buildHook('root')) // [1]
app.register(async function pluginOne(pluginInstance, opts)
{
  pluginInstance.addHook('onRoute', buildHook('pluginOne'))
    // [2]
  pluginInstance.get('/one', async () => 'one')
})
app.register(async function pluginTwo(pluginInstance, opts {
  pluginInstance.addHook('onRoute', buildHook('pluginTwo'))
    // [3]
  pluginInstance.get('/two', async () => 'two')
  pluginInstance.register(async function pluginThree(
  subPlugin, opts) {
    subPlugin.addHook('onRoute', buildHook('pluginThree'))
      // [4]
    subPlugin.get('/threee', async () => 'three')
  })
})
```

```
function buildHook(id) {
  return function hook(routeOptions) {
    console.log(`onRoute ${id} called from ${routeOptions
      .path}`)
  }
}
```

Running the preceding code will execute [2] and [4] just one time, because inside pluginOne and pluginThree, only one route has been registered, and each plugin has registered only one hook. The onRoute hook [1] is executed three times, instead. This happens because it has been added to the app instance, which is the parent scope of all the plugins. For this reason, the root hook will listen to the events of its context and to the children's ones.

This feature comes with an endless list of benefits that you will "get to know" through this book. To better explain the bigger advantage of this feature, imagine every plugin as an isolated box that may contain other boxes, and so on, where the Root application instance is the primary container of all the plugins. The previous code can be schematized in this diagram:

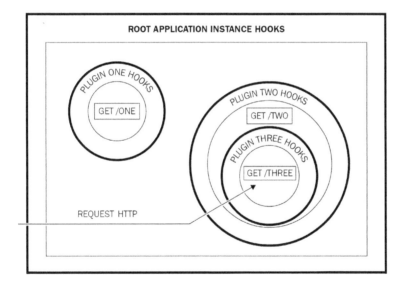

Figure 1.2 – Encapsulated contexts

The request is routed to the right endpoint (the square in the diagram), and it will trigger all the hooks registered on each plugin instance that include the destination handler.

Every box is self-contained, and it won't affect the behavior of its other siblings, thus giving you the certainty that no issue affects parts of the application other than the one where it occurred. Furthermore, the system only executes the hook functions your routes need! This allows you and your team to work on different parts of the application without affecting each other or causing side effects. Furthermore, the isolation will give you a lot more control over what is happening at your endpoints. Just to give you an example: you can add a dedicated database connection for hot-paths in your code base without extra effort!

This plugin example has shown more clearly the Fastify plugin system in action. It should help you understand the difference between a *root application instance* and *plugin instances*. You now have an idea of how powerful the plugin system is and of the benefits it implements by design:

- **Encapsulation**: All the hooks, plugins, and decorators added to the plugin are binded to the plugin context
- **Isolation**: Every plugin instance is self-contained and doesn't modify sibling plugins
- **Inheritance**: A plugin inherits the configuration of the parent plugin

The plugin system will be discussed in depth in *Chapter 2*.

Now, we are ready to explore how to manage the different configuration types a Fastify application needs to work correctly.

Understanding configuration types

In Fastify, we must consider splitting the configuration into three types to better organize our application:

- **Server options**: Provide the settings for the Fastify framework to start and support your application. We have presented them before when describing how to instantiate the server instance in the *The root application instance* section.
- **Plugin configuration**: Provides all the parameters to configure your plugins or the community plugins.
- **Application configuration**: Defines your endpoint settings.

This can be implemented with a configuration loader function:

```
const environment = process.env.NODE_ENV // [1]
async function start () {
  const config = await staticConfigLoader(environment) // 2
  const app = fastify(config.serverOptions.factory)
  app.register(plugin, config.pluginOptions.fooBar)
  app.register(plugin, { // [3]
    bar: function () {
      return config.pluginOptions ? 42 : -42
```

```
    }
  })
  await app.listen(config.serverOptions.listen)
  async function staticConfigLoader (env) {
    return { // [4]
      env,
      serverOptions: getServerConfig(),
      pluginOptions: {},
      applicationOptions: {}
    }
  }
}
start()
```

This example shows the key points of a configuration loader:

1. It must accept the environment as input. This will be fundamental during the test writing.

2. It should be an async function: you will load settings from a different source that needs I/O.

3. It must manage primitive types exclusively.

4. It can be split into three main objects for clarity.

A plugin's configuration often needs an input parameter that is not a primitive type-like function. This would be part of the code flow since a function acts based on input strings such as passwords, URLs, and so on.

This quick introduction shows you the logic we need to take into consideration when we build more complex code. This separation helps us to think about how to better split our configuration files. We will read a complete example in *Chapter 6*.

Now, we can configure and start our Fastify server; it is time to turn it off.

Shutting down the application

Up until now, we have killed our server by pressing the *Ctrl + C* or *Cmd+ C* keys. This shortcut sends a SIGINT interrupt to the Node.js process, which will cause an unconditional termination. If we don't manage this behavior, a running HTTP request may be interrupted, causing possible data loss or introducing inconsistencies in our application.

To ensure you close the server gracefully, you must call the root instance's close method:

```
process.once('SIGINT', function closeApplication() {
  app.close(function closeComplete(err) {
    if (err) {
      app.log.error(err, 'error turning off')
    } else {
```

```
        app.log.info('bye bye')
      }
    })
  })
```

Adding this signaling handle will prevent the kill of the server, thus allowing the complete execution of the requests and preventing new HTTP requests from being accepted. New requests will receive the `HTTP Status 503 - Service Unavailable` error while the application is in the closing phase.

Calling the `close` method will trigger the `onClose` hook too. All the plugins that are listening for this event will receive it at the beginning of the shutdown process, as a database plugin will close the connection.

Fastify guarantees that the `onClose` hooks will be executed once, even when the server's `close` method is called multiple times. Note that the `close` callback function will be run at every call instead.

Our implementation, unfortunately, is not enough to cover all the use cases one application may face. If the plugins don't resolve the `onClose` hook, due to a bug or a starving connection, our server will become a zombie that will wait forever to close gracefully. For this reason, we need to develop a maximum time span, after which the application must stop. So, let's analyze an example of force close using the Fastify async interface:

```
process.once('SIGINT', async function closeApplication() {
  const tenSeconds = 10_000
  const timeout = setTimeout(function forceClose() {
    app.log.error('force closing server')
    process.exit(1)
  }, tenSeconds)
  timeout.unref()

  try {
    await app.close()
    app.log.info('bye bye')
  } catch (err) {
    app.log.error(err, 'the app had trouble turning off')
  }
})
```

We have set a timeout timer in the previous code that doesn't keep the Node.js event loop active, thanks to the `unref` call. If the close callback doesn't execute in 10 seconds, the process will exit with a nonzero result. This pattern is implemented in many plugins built by Fastify's community that you can check out on the *Ecosystem* page at `https://www.fastify.io/ecosystem/`.

Turning off a server could be challenging, but Fastify provides a set of features that help us to avoid losing data and complete all the application's pending tasks. We have seen how to deal with it through a pattern that guarantees to stop the server in a reasonable time. Looking at the community plugins is a good way to learn how to search for an external plugin that implements the pattern and provides us with this feature, without having to re-implement it ourselves.

Summary

This first chapter offered a complete overview of Fastify's framework. It touched on all the essential features it offers, thus allowing you to start "playing" with the applications.

So far, you have learned how to instantiate a server, add routes, and turn it on and off gracefully. You have seen the basic logger configuration and have learned how to use it. However, more importantly, you have understood the basic concepts behind Fastify and its components. You had a quick insight into request hooks and more comprehensive examples regarding application hooks. Lastly, you were presented with a simple outline of the lifecycle that controls the execution of the code.

Now that you are confident with the syntax, get ready to explore the details of this great framework!

In the next chapter, we will expand the Fastify booting process and discover new possibilities that are going to make our daily job much easier. Moreover, we will explore the plugin system in depth in order to become proficient and start building our first plugins.

2

The Plugin System and the Boot Process

A Fastify plugin is an essential tool at the disposal of a developer. Every functionality except the root instance of the server should be wrapped in a plugin. Plugins are the key to reusability, code sharing, and achieving proper encapsulation between Fastify instances.

Fastify's root instance will load all registered plugins asynchronously following the registration order during the boot sequence. Furthermore, a plugin can depend on others, and Fastify checks these dependencies and exits the boot sequence with an error if it finds missing ones.

This chapter starts with the declaration of a simple plugin and then, step by step, adds more layers to it. We will learn why the `options` parameter is crucial and how the Fastify register method uses it during the boot sequence. The final goal is to understand how plugins interact with each other thanks to encapsulation.

To understand this challenging topic, we will introduce and learn about some core concepts:

- What is a plugin?
- The `options` parameter
- Encapsulation
- The boot sequence
- Handling boot and plugin errors

Technical requirements

To follow this chapter, you will need the following:

- A text editor, such as VS Code

- A working Node.js v18 installation

- Access to a shell, such as Bash or CMD

All the code examples in this chapter can be found on GitHub at `https://github.com/PacktPublishing/Accelerating-Server-Side-Development-with-Fastify/tree/main/Chapter%202`.

What is a plugin?

A Fastify **plugin** is a component that allows developers to extend and add functionalities to their server applications. Some of the most common use cases for developing a plugin are handling a database connection or extending default capabilities – for example, to request parsing or response serialization.

Thanks to their unique properties, plugins are the basic building blocks of our application. Some of the most prominent properties are the following:

- A plugin can register other plugins inside it.

- A plugin creates, by default, a new scope that inherits from the parent. This behavior also applies to its children and so on, although using the parent's context is still possible.

- A plugin can receive an `options` parameter that can be used to control its behavior, construction, and reusability.

- A plugin can define scoped and prefixed routes, making it the perfect router.

At this point, it should be clear that where other frameworks have different entities, such as middleware, routers, and plugins, Fastify has only plugins. So, even if Fastify plugins are notoriously hard to master, we can reuse our knowledge for almost everything once we understand them!

We need to start our journey into the plugin world from the beginning. In the following two sections, we will learn how to declare and use our first plugin.

Creating our first plugin

Let's see how to develop our first dummy plugin. To be able to test it, we need to register it to a root instance. Since we focus on plugins here, we will keep our Fastify server as simple as possible, reusing the basic one we saw in *Chapter 1* and making only one minor change. In index.cjs, we will declare our plugin inline, and then we will see how to use separate files for different plugins:

```
const Fastify = require('fastify')
const app = Fastify({ logger: true }) // [1]
app.register(async function (fastify, opts) { // [2]
    app.log.info('Registering my first plugin.')
})
app.ready() // [3]
  .then(() => { app.log.info('All plugins are now
    registered!')
  })
```

After creating a Fastify root instance ([1]), we add our first plugin. We achieve this by passing a plugin definition function as the first argument ([2]) to register. This function receives a new Fastify instance, inheriting everything the root instance has until this point and an options argument.

> **The options argument**
>
> We are not using the options argument at the moment. However, in the next section, we will learn about its importance and how to use it.

Finally, we call the ready method ([3]). This function starts the boot sequence and returns Promise, which will be settled when all plugins have been loaded. For the moment, we don't need a listening server in our examples, so it is acceptable to call ready instead. Moreover, listen internally awaits for the .ready() event to be dispatched anyway.

> **The boot sequence**
>
> The Fastify boot sequence is a series of operations that the Fastify primary instance performs to load all of the plugins, hooks, and decorators. If no errors are encountered, the server will start and listen on the provided port. We will learn more about it in a separate section.

Let's run the previous snippet in the terminal and look at the output:

```
$ node index.cjs
{"level":30,"time":1621855819132,"pid":5612,"hostname":"my.
local","msg":"Registering my first plugin."}
{"level":30,"time":1621855819133,"pid":5612,"hostname":"my.
local","msg":"All plugins are now registered!"}
```

Let's break down the execution of our snippet:

1. The Fastify instance is created with the logger enabled.

2. The Fastify root instance registers our first plugin, and the code inside the plugin function is executed.

3. The `Promise` instance returned by the ready method is resolved after the ready event is dispatched.

This Node.js process exits without any errors; this happens because we used `ready` instead of the `listen` method. The declaration method we just saw is only one of the two that are possible. In the next section, we will look at the other one, since many online examples use it.

The alternative plugin function signature

As with almost every API it exposes, Fastify has two alternative ways to declare a plugin function. The only difference is the presence, or absence, of the third callback function argument, usually called `done`. Following the good old Node.js callback pattern, this function has to be invoked to notify that the loading of the current plugin is done with no errors.

On the other hand, if an error occurs during the loading, `done` can receive an optional `err` argument that can be used to interrupt the boot sequence. This same thing happens in the promise world – if the promise is resolved, the plugin is loaded; if the promise is rejected, the rejection will be passed up to the root instance, and the boot process will terminate.

Let's see the `callbacks.cjs` snippet that uses the callback style plugin definition:

```
//...
app.register(function noError(fastify, opts, done) {
app.log.info('Registering my first plugin.')
    // we need to call done explicitly to let fastify go to
        the next plugin
    done() // [1]
})
app.register(function (fastify, opts, done) {
app.log.info('Registering my second plugin.')
try {
    throw new Error('Something bad happened!')
    done() // [2]
} catch (err) {
    done(err) // [3]
}
})
```

The first plugin loads without any errors, thanks to the done call with no arguments passed ([1]). On the other hand, the second one throws before done() ([2]) is called. Here, we can see the importance of error catching and calling done(err) ([3]); without this call, Fastify would think no errors happened during the registering process, continuing the boot sequence!

No matter which style you use, plugins are always asynchronous functions, and every pattern has its error handling. It is essential not to mix the two styles and to be consistent when choosing one or another:

- Calling done if the callback style is used
- Resolving/rejecting the promise if promise-based is the flavor we chose

We will come back to this argument again in the *Boot errors* section, where we will see what happens if we misuse promises and callbacks.

> **Promise-based examples**
>
> This book uses promise-based signatures since they are easier to follow, thanks to the async/await keywords.

In this section, we learned the two different methods to declare our first plugin and register it with a Fastify instance. However, we have just scratched the surface of Fastify plugins. In the next section, we will look at the options parameter and how it comes in handy when dealing with sharing functionalities.

Exploring the options parameter

This section is a deep look at the optional options parameter and how we can develop reusable plugins. A plugin declaration function isn't anything more than a factory function that, instead of returning some new entity as factory functions usually do, adds behaviors to a Fastify instance. If we look at it like this, we can think about the options parameter as our constructor's arguments.

Firstly, let's recall the plugin declaration function signature:

```
async function myPlugin(fastify, [options])
```

How can we pass our custom arguments to the options parameter? It turns out that the register method has a second parameter too. So, the object we use as the argument will be passed by Fastify as Plugin's options parameter:

```
app.register(myPlugin, { first: 'option'})
```

Now, inside the myPlugin function, we can access this value simply using options.first.

It is worth mentioning that Fastify reserves three specific options that have a special meaning:

- `prefix`
- `logLevel`
- `logSerializers`

> **Logger options**
>
> We will cover `logLevel` and `logSerializer` in a dedicated chapter. Here, we will focus only on `prefix` and custom options.

Bear in mind that, in the future, more reserved options might be added. Consequently, developers should always consider using a namespace to avoid future collisions, even if it is not mandatory. We can see an example in the `options-namespacing.cjs` snippet:

```
app.register(async function myPlugin(fastify, options) {
  console.log(options.myplugin.first)
}, {
  prefix: 'v1',
  myPlugin: {
    first: 'custom option',
  }
})
```

Instead of adding our custom properties at the top level of the `options` parameter, we will group them in a custom key, lowering the chances for future name collisions. Passing an object is fine in most cases, but sometimes, we need more flexibility.

In the next section, we will learn more about the `options` parameter type and leverage it to do more complex stuff.

The options parameter type

So far, we have seen that the `options` parameter is an object with some reserved and custom properties. But `options` can also be a function that returns an object. If a function is passed, Fastify will invoke it and pass the returned object as the `options` parameter to the plugin. To better understand this, in the `options-function.cjs` snippet, we will rewrite the previous example using the function instead:

```
app.register(async function myPlugin(fastify, options) {
  app.log.info(options.myplugin.first) // option
}, function options(parent) { // [1]
  return ({
```

```
        prefix: '1',
        myPlugin: {
          first: 'option',
        }
    })
  })
```

At first glance, it shouldn't be clear why we have this alternative type for options, but looking at the signature points us in the right direction – it receives Fastify's parent instance as the only argument.

Looking at the `options-function-parent.cjs` example should clarify how we can access the parent options:

```
const Fastify = require('fastify')
function options(parent) {
  return ({
    prefix: 'v1',
    myPlugin: {
      first: parent.mySpecialProp, // [2]
    }
  })
}
const app = Fastify({ logger: true })
app.decorate('mySpecialProp', 'root prop') // [1]
app.register(async function myPlugin(fastify, options) {
  app.log.info(options.myplugin.first) // 'root prop'
}, options)
app.ready()
```

First, we decorate the root instance with a custom property ([1]), and then we pass it as a value to our plugin ([2]). In a real-world scenario, `mySpecialProp` could be a database connection, any value that depends on the environment, or even something another plugin has added.

The prefix option

At the beginning of this chapter, we learned that we can define routes inside a plugin. The `prefix` option comes in handy here because it allows us to add a namespace to our route declarations. There are several use cases for this, and we will see most of them in the more advanced chapters, but it is worth mentioning a couple of them here:

- Maintaining different versions of our APIs

- Reusing the same plugin and routes definition for various applications, giving another mount point each time

The `users-router.cjs` snippet will help us understand this parameter better. First of all, we define a plugin in a separate file and export it:

```
module.exports = async function usersRouter(fastify, _) {
  fastify.register(
    async function routes(child, _options) { // [1]
      child.get('/', async (_request, reply) => {
        reply.send(child.users)
      })
      child.post('/', async (request, reply) => { // [2]
        const newUser = request.body
        child.users.push(newUser)
        reply.send(newUser)
      })
    },
    { prefix: 'users' }, // [3]
  )
}
```

> **Route declaration**
>
> Since we are focusing on plugins here and trying to keep the examples as short as possible, this section uses the "shorthand declaration" style for routes. Moreover, schemas and some other crucial options are missing too. We will see through this book that there are much better, formally correct, and complete options for route declarations.

We define two routes; the first one returns all of the elements in our collection ([1]), and the second one enables us to add entries to the user's array ([2]). Since we don't want to add complexity to the discussion, we use an array as our data source; it is defined on the root Fastify instance, as we will learn in the following snippet. In a real-world scenario, this would be, of course, some kind of database access. Finally, at [3], we prefix all of the routes we define with the user's namespace in the old RESTful fashion.

Now that we have defined the namespace, we can import and add this **router** to the root instance in `index-with-router.cjs`. We can also use the **prefix** option to give a unique namespace to our routes and handle API versioning:

```
const Fastify = require('fastify')
const usersRouter = require('./users-router.cjs')

const app = Fastify()
app.decorate('users', [ // [1]
  {
```

```
      name: 'Sam',
      age: 23,
    },
    {
      name: 'Daphne',
      age: 21,
    },
  ])

app.register(usersRouter, { prefix: 'v1' }) // [2]
app.register(
  async function usersRouterV2(fastify, options) { // [3]
    fastify.register(usersRouter) // [4]
    fastify.delete('/users/:name', (request, reply) => { //
    [5]
      const userIndex = fastify.users.findIndex(
        user => user.name === request.params.name,
      )
      fastify.users.splice(userIndex, 1)
      reply.send()
    })
  },
  { prefix: 'v2' },
)

app.ready()
  .then(() => { console.log(app.printRoutes()) }) // [6]
```

First of all, we decorate the Fastify root instance with the `users` property ([1]); as previously, this will act as our database for this example. On [2], we register our user's router with the `v1` prefix. Then, we register a new inline-declared plugin ([3]), using the `v2` namespace (every route added in this plugin will have the `v2` namespace). On [4], we register the same user's routes for a second time, and we also add a newly declared `delete` route ([5]).

> **The printRoutes method**
>
> This method can be helpful during development. If we are not sure of the full path of our routes, it prints all of them for us!

Thanks to **[6]**, we can discover all the routes we mounted:

```
$ node users-router-index.cjs
└── /
```

```
├── v
│   ├── 1
│   │   └── /users (GET)
│   │       /users (POST)
│   │       └── / (GET) # [1]
│   │           / (POST)
│   └── 2
│       └── /users (GET)
│           /users (POST)
│           └── / (GET) # [1]
│               / (POST)
└── v2/users/:name (DELETE)
```

Indeed, prefixing route definitions is a compelling feature. It allows us to reuse the same route declarations more than once. It is one of the crucial elements of the reusability of the Fastify plugins. In our example, we have just two levels of prefix nesting, but there are no limits in practice. We avoid code duplication, using the same GET and POST definitions twice and adding only one new DELETE route to the same user's namespace when needed.

This section covered how to use the options parameter to achieve better plugin reusability and control its registration on the Fastify instance. This parameter has some reserved properties used to tell Fastify how to handle the registering plugin. Furthermore, we can add as many properties as needed by the plugin, knowing that Fastify will pass them during the registration phase.

Since we have already used **encapsulation** in this section without even knowing it, it will be the topic of the next section.

Understanding encapsulation

So far, we've written a few plugins. We are pretty confident about how they are structured and what arguments a plugin receives. We still need to discuss one missing thing about them – the concept of encapsulation.

Let's recall the plugin function definition signature:

```
async function myPlugin(fastify, options)
```

As we know at this point, the first parameter is a Fastify instance. This instance is a newly created one that inherits from the outside scope. Let's suppose something has been added to the root instance, for example, using a decorator. In that case, it will be attached to the plugin's Fastify instance, and it will be usable as if it is defined inside the current plugin.

The opposite isn't true, though. If we add functionalities inside a plugin, those things will be visible only in the current plugin's context.

> **Context versus scope**
>
> Firstly, let's take a look at the definitions of both terms. The *context* indicates the current value of the implicit 'this' method variable. The *scope* is a set of rules that manages the visibility of a variable from a function point of view. In the Fastify community, these two terms are used interchangeably and refer to the Fastify instance we are currently working with. For this reason, in this book, we will use both words, meaning the same thing.

Let's take a look at the example in encapsulation.cjs:

```
const Fastify = require('fastify')

const app = Fastify({ logger: true })

app.decorate('root', 'hello from the root instance.') // [1]

app.register(async function myPlugin(fastify, _options) {
  console.log('myPlugin -- ', fastify.root)
  fastify.decorate('myPlugin', 'hello from myPlugin.') //
  [2]
  console.log('myPlugin -- ', fastify.myPlugin)
})

app.ready()
  .then(() => {
    console.log('root -- ', app.root) // [3]
    console.log('root -- ', app.myPlugin)
})
```

Running this snippet will produce this output:

```
$ node encapsulation.cjs
myPlugin --   hello from the root instance.
myPlugin --   hello from myPlugin.
root --   hello from the root instance.
root --   undefined
```

Firstly, we decorate the root instance ([1]), adding a string to it. Then, inside myPlugin, we print the root decorated value and add a new property to the Fastify instance. In the plugin definition body, we log both values in the console to ensure they are set ([2]). Finally, we can see that after the Fastify application is ready, at the root level, we can only access the value we added outside of our plugin ([3]).

But what happened here? In both cases, we used the .decorate method to add our value to the instance. Why are both values visible in myPlugin but only the root one visible at the top level? This is the intended behavior, and it happens thanks to encapsulation – Fastify creates a new context every time it enters a new plugin. We call these new contexts **child contexts**. A child context inherits only from the parent contexts, and everything added inside a child context will not be visible to its parent or its siblings' contexts. The parent-child annidation level is infinite, and we can have contexts that are children to their parents and parents to their children.

The entities that are affected by scoping are:

- **Decorators**
- **Hooks**
- **Plugins**
- **Routes**

As we can see, since **routes** are affected by context, we already used encapsulation in the previous section, even if we didn't know it at the time. We registered the same route on the same root instance twice, but with different prefixes.

In real-world applications, more complex scenarios with several child and grandchild contexts are widespread. We can use the following diagram to examine a more complex example:

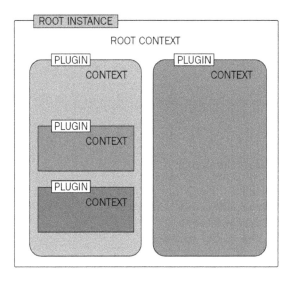

Figure 2.1: An example of a complex plugin hierarchy

In *Figure 2.1*, we can see a rather complex scenario. We have a root Fastify instance that registers two root plugins. Every root plugin creates a new child context where we can again declare and register as many plugins as we want. That's it – we can have infinite nesting for our plugins, and every level depth will create a new encapsulated context.

However, Fastify leaves complete control over encapsulation for the developer, and we will see how to control it in the next section.

Handling the context

Until now, the context we relied upon was based on the default Fastify behavior. It works most of the time, but there are some cases where we need more flexibility. If we need to share context between siblings or alter a parent context, we can still do it. This comes in handy for more complex plugins, such as ones that handle database connections.

We have a couple of tools at our disposal:

- The `skip-override` hidden property
- The `fastify-plugin` package

We will start using `skip-override` and then move on to `fastify-plugin`; even though they can be used to achieve the same result, the latter has additional features.

Here, we will use the same example we used before but now add the `skip-override` hidden property to ensure we can access the decorated variable at the top-level scope. The `skip-override.cjs` snippet will help us understand its usage:

```
const Fastify = require('fastify')

async function myPlugin(fastify, _options) {
  console.log('myPlugin -- ', fastify.root)
  fastify.decorate('myPlugin', 'hello from myPlugin.')
// [2]
  console.log('myPlugin -- ', fastify.myPlugin)
}

myPlugin[Symbol.for('skip-override')] = true // [1]

const app = Fastify({ logger: true })
app.decorate('root', 'hello from the root instance.')
app.register(myPlugin)

app.ready()
  .then(() => {
    console.log('root -- ', app.root)
console.log('root -- ', app.myPlugin) // [3]
})
```

There is only one major change in this code snippet from the previous one, which is that we use `Symbol.for('skip-override')` to prevent Fastify from creating a new context ([1]). This alone is enough to have the root decorated with the `fastify.myPlugin` variable ([2]). We can see that the decorator is also accessible from the outer scope ([3]); here is the output:

```
$ node skip-override.cjs
myPlugin --  hello from the root instance.
myPlugin --  hello from myPlugin.
root --  hello from the root instance.
root --  hello from myPlugin.
```

> **Symbol.for**
>
> Instead of using just the `skip-override` property as a string, Fastify uses `Symbol.for` to hide it and avoid name collisions. When a symbol is created, it is added to a runtime-wide symbol registry. Upon the invocation, the `.for` method checks whether the symbol we are trying to access is already present. If not, it first creates it and then returns it. More on symbols can be found at `https://developer.mozilla.org/en-US/docs/Web/JavaScript/Reference/Global_Objects/Symbol`.

fastify-plugin

Using `skip-override` is perfectly fine, but there is a better way to control encapsulation behavior. Like many things in the Fastify world, the `fastify-plugin` module is nothing more than a function that wraps a plugin, adds some metadata to it, and returns it.

The main features included with `fastify-plugin` are the following:

- Adding the `skip-override` property for us
- Providing the name for the plugin, if no explicit one is passed
- Checking the minimum version of Fastify
- Attaching the provided custom metadata to the returned plugin

We should use `fastify-plugin` every time we have a plugin that needs to add behavior to the parent context. To be able to use it, we first need to install it from the `npmjs.com` registry:

```
$ npm i fastify-plugin
```

Its signature resembles the `.register` method. It takes two parameters:

- The exact plugin function definition, as we have already seen
- An optional `options` object with four optional properties:

- name: We can use this string property to give a name to our plugin. Giving a name to our plugin is fundamental because it allows us to pass it into the dependencies array, as we will see soon. Moreover, if something unexpected happens during the boot, Fastify will use this property for better stack traces.

- fastify: We can use this string property to check the minimum Fastify version that our plugin needs to work correctly. This property accepts as a value any valid **SemVer** range.

- decorators: We can use this object to ensure the parent instance has been decorated with properties that we use inside our plugin.

- dependencies: If our plugin depends on other plugins' functionalities, we can use this array of strings to check that all dependencies are met – the values are the plugin name. We will take a deeper look at this property in the next chapter, since it is related to the boot process.

SemVer

SemVer stands for **Semantic Versioning**, and its purpose is to help developers manage their dependencies. It is composed of three numbers separated by two dots, following the MAJOR. MINOR.PATCH schema. If any breaking changes are added to the code, then the MAJOR number needs to be increased. On the other hand, if new features are added but no breaking changes are introduced, then the MINOR number should be increased. Finally, if all of the changes are done only to fix bugs, then the PATCH number is bumped.

The fp naming convention

In the Fastify community, it is common to import the fastify-plugin package as fp because it is a short yet meaningful variable name. In this book, we will use this convention every time we deal with it.

Let's take a look at the fp-myplugin.cjs example that uses the optional metadata properties:

```
const fp = require('fastify-plugin')
async function myPlugin(fastify, _options) {
  console.log('myPlugin decorates the parent instance.')
  fastify.decorate('myPlugin', 'hello from myPlugin.')
}
module.exports = fp(myPlugin, { // [1]
  name: 'myPlugin', // [2]
  fastify: '4.x', // [3]
  decorators: { fastify: ['root'] }, // [4]
})
```

At **[1]**, we pass `myPlugin` as the first argument and export the wrapped plugin as the default export. The second argument is the options objects:

- We give an explicit name to our plugin (**[2]**).

- We set the minimum Fastify version to `4.x` (**[3]**).

- The `decorators` property accepts the keys of the entities that can be decorated – `fastify`, `request`, and `reply`. Here, we check whether the Fastify parent instance has the `root` property set (**[4]**).

The `fastify-plugin` attaches the `options` object to our plugin in the unique hidden property called `Symbol.for('plugin-meta')`. Fastify will look for this property during the plugin registration, and if found, it will act accordingly to its content.

In the `fp-myplugin-index.cjs` snippet, we import and register our plugin, checking the different outcomes:

```
const Fastify = require('fastify')
const myPlugin = require('./fp-myplugin.cjs')

const app = Fastify({ logger: true })
app.decorate('root', 'hello from the root instance.') //
[1]
app.register(myPlugin) // [2]

app.ready()
  .then(() => {
    console.log('root -- ', app.root)
console.log('root -- ', app.myPlugin) //[3]
})
```

First, we decorate the root fastify instance with a string property (**[1]**). Then, we register our plugin (**[2]**). Remember that we specified the `root` property as mandatory inside the metadata of `fastify-plugin`. Before registering `myPlugin`, Fastify checks whether the property is declared on the parent context, and since it is there, it goes on with the boot process. Finally, since `fastify-plugin` adds the `skip-override` property for us, we can access the `myPlugin` property in the root scope with no issues (**[3]**). Let's take a look at the output of this snippet:

```
$ node fp-myplugin-index.cjs
myPlugin decorates the parent instance.
root --  hello from the root instance.
root --  hello from myPlugin.
```

Everything works as expected!

Now, looking at `fp-myplugin-index-missing-root.cjs`, we can check what happens if the `root` decorator is missing from the root instance, as it was declared in `fp-myplugin-index.cjs` at [1], as shown previously:

```
const Fastify = require('fastify')
const myPlugin = require('./fp-myplugin.cjs')

const app = Fastify({ logger: true })
app.register(myPlugin)

app.ready()
  .then(() => {
    console.log('root -- ', app.root)
console.log('root -- ', app.myPlugin)
})
```

Running this file will throw and abort the boot process:

```
$ node fp-myplugin-index-missing-root.cjs
/node_modules/fastify/lib/pluginUtils.js:98
      throw new FST_ERR_PLUGIN_NOT_PRESENT_IN_INSTANCE(decorator,
withPluginName, instance)
            ^

FastifyError [Error]: The decorator 'root' required by 'myPlugin' is
not present in Fastify
    at /node_modules/fastify/lib/pluginUtils.js:98:13
    at Array.forEach (<anonymous>)
    at _checkDecorators (/node_modules/fastify/lib/pluginUtils.
js:95:14)
    at Object.checkDecorators (/node_modules/fastify/lib/pluginUtils.
js:81:27)
    at Object.registerPlugin (/node_modules/fastify/lib/pluginUtils.
js:137:19)
    at Boot.override (/node_modules/fastify/lib/pluginOverride.
js:28:57)
    at Plugin.exec (/node_modules/avvio/plugin.js:79:33)
    at Boot.loadPlugin (/node_modules/avvio/plugin.js:272:10)
    at processTicksAndRejections (node:internal/process/task_
queues:83:21) {
  code: 'FST_ERR_PLUGIN_NOT_PRESENT_IN_INSTANCE',
  statusCode: 500
}
```

We can see that Fastify used the `myPlugin` name in the `FST_ERR_PLUGIN_NOT_PRESENT_IN_INSTANCE` error, helping us to understand the issue. This is very helpful when dealing with dozens of registered plugins.

> **The name property**
>
> We saw that the `fastify-plugin options` parameter is optional, and so is its `name` property. But what happens if we don't pass it? It turns out that a name will be generated and attached to the plugin. If our plugin is a named function (it contains the name property), it will be used as the plugin name. Otherwise, the filename is the next candidate. In both cases, a standard part will be appended – `auto-{N}`. The N is an auto-incremented number that starts from 0. The appended part is needed to avoid naming collisions since the developer does not provide these names, and Fastify doesn't want to block the boot process for unintended collisions. It is important to remember that giving an explicit name to our plugins is considered a best practice.

This section covered one of the core yet most difficult concepts in Fastify – **encapsulation**. We learned how Fastify provides default behavior that suits the most common cases but gives full power to the developer whenever needed. In addition, tools such as `skip-override` and `fastify-plugin` are fundamental when dealing with more complex scenarios, where control over the context is crucial.

But how does Fastify know the correct order of plugin registration? Is it even a deterministic process? We will discover this and more about the boot sequence in the next section.

Exploring the boot sequence

We learned in the previous section that a plugin is just an asynchronous function with well-defined parameters. We've also seen how Fastify plugins are the core entity we use to add features and functionalities to our applications. In this section, we will learn what the boot sequence is, how plugins interact with each other, and how Fastify ensures that all the developer's constraints are met, before running the HTTP server.

Firstly, it is essential to say that the Fastify boot sequence is asynchronous too. Fastify loads every plugin added with the `register` method, one by one, respecting the order of the registration. Fastify starts this process only after `.listen()` or `.ready()` are called. After that, it waits for all promises to be settled (or for all completed callbacks to be called, if the callback style is used), and then it emits the ready event. If we have already got this far, we can be sure that our application is up and running and ready to receive incoming requests.

> **Avvio**
>
> The boot process is baked by `Avvio`, a library that can also be used standalone. `Avvio` handles all the complexities concerning the asynchronous boot – error handling, loading order, and dispatching the ready event, enabling developers to ensure that the application is started after everything is loaded without any errors.

This somehow overshadowed and underrated feature is, in reality, one of the most powerful ones. Having an asynchronous boot process brings to the table some key benefits:

- After the `listen`/`ready` promise is resolved, we are sure that all plugins are loaded without errors and the boot sequence is over

- It allows us to always register our plugins in a deterministic way, ensuring the loading and closing orders

- If an error is thrown during the plugin registration, it will interrupt the boot sequence, enabling developers to act accordingly

Even if Fastify's boot process is very versatile, it handles almost everything out of the box. The exposed API is small – there are just two methods! The first one is the `register` method, which we have already used many times. The other one is `after`, and as we will see, it is rarely used because `register` integrates its functionalities for most of the use cases. Let's take a deeper look at them.

---

thenable

Several Fastify instance methods return `thenable`, an object that has the `then()` method. The most important property of `thenable` is that the promise chains and `async/await` work fine. Using `thenable` instead of a real promise has one main advantage – it enables the object to be simultaneously awaited and chained to other calls in a fluent API fashion.

The register instance method

At this point, we almost know everything about this core method. Let's start with its signature:

```
.register(async function plugin(instance, pluginOptions), options)
```

The `options` parameter is fundamental to passing custom options to our plugins during the registration, and it is the gateway to plugin reusability. For example, we can register the same plugin that deals with the database with a different connection string or, as we saw, use another prefix option to register the same handler on different route paths. Moreover, if we don't use the `fastify-plugin` module or the `skip-override` hidden property, `register` creates a new context, isolating whatever our plugin does.

But what is the `register` return value? It is a `thenable` Fastify instance, and it brings two significant benefits to the table:

- It adds the ability to chain instance method calls.

- If needed, we can use `await` when calling `register`

First, we will take a look at Fastify instance method chaining in `register-chain.cjs`:

```
const Fastify = require('fastify')

// [1]
async function plugin1(fastify, options) {
  app.log.info('plugin1')
}
async function plugin2(fastify, options) {
  app.log.info('plugin2')
}
async function plugin3(fastify, options) {
  app.log.info('plugin3')
}

const app = Fastify({ logger: true })

app // [2]
  .register(plugin1)
  .register(plugin2)

app.register(plugin3) // [3]

app.ready()
  .then(() => {
    console.log('app ready')
  })
```

We define three dummy plugins that do only one thing when registered – they log their function name ([**1**]). Then, at [**2**], we register both of them using method chaining. The first `.register(plugin1)` invocation returns the Fastify instance, allowing the subsequent call, `.register(plugin2)`. We can use method chaining with the majority of instance methods. We can break the invocation chain and call the register directly on the instance ([**3**]).

Awaiting register

There is one more thing to say about the register method. We can use the `await` operator after every `register` call to wait for the registration to be done. The `await` operator can be used after any `register` call to wait for every plugin added up to that point to load, as we can see in the following `await-register.cjs` snippet:

```
const Fastify = require('fastify')
const fp = require('fastify-plugin')

async function boot () {
```

```
  // [1]
  async function plugin1 (fastify, options) {
    fastify.decorate('plugin1Decorator', 'plugin1')
  }
  async function plugin2 (fastify, options) {
    fastify.decorate('plugin2Decorator', 'plugin2')
  }
  async function plugin3 (fastify, options) {
    fastify.decorate('plugin3Decorator', 'plugin3')
  }

  const app = Fastify({ logger: true })
  await app // [2]
    .register(fp(plugin1))
    .register(fp(plugin2))

  console.log('plugin1Decorator',
    app.hasDecorator('plugin1Decorator'))
  console.log('plugin2Decorator',
    app.hasDecorator('plugin2Decorator'))

  app.register(fp(plugin3)) // [3]
  console.log('plugin3Decorator',
    app.hasDecorator('plugin3Decorator'))

  await app.ready()
  console.log('app ready')
  console.log('plugin3Decorator',
    app.hasDecorator('plugin3Decorator'))
}
boot()
```

To understand what is going on, we will run this snippet:

```
$ node register-await.mjs
plugin1Decorator true # [4]
plugin2Decorator true
plugin3Decorator false # [5]
app ready
plugin3Decorator true # [6]
```

Since this example is quite long and complex, let's break it into smaller chunks:

- We declare three plugins [1], each adding one decorator to the Fastify instance

- We use `fastify-plugin` to decorate the root instance and to register these plugins ([2] and [3]), as we learned in the previous section

- Even if our plugins are identical, we can see that the results of `console.log` are different; the two plugins registered with the `await` operator ([2]) have already decorated the root instance ([4])

- Conversely, the last one registered without `await` ([3]) adds its decorator only after the ready event promise has been resolved ([5] and [6])

At this point, it should be clear that if we also want the third decorator to be available before the application is fully ready, it is sufficient to add the `await` operator on [3]!

As a final note, we can say that we don't need to wait when registering our plugins for most cases. The only exception might be when we need access to something that the plugin did during the loading. For example, we have a plugin that connects to an external source, and we want to be sure that it is connected before going on with the boot sequence.

The after instance method

The function argument of the `after` method is called automatically by Fastify when all plugins added until that point have finished loading. As the only parameter, it defines a callback function with an optional error argument:

```
.after(function (err) {})
```

If we don't pass the callback, then `after` will return a `thenable` object that can be awaited. In this version, awaiting `after` can be replaced by just awaiting `register`; the behavior is the same. Because of that, in the `after.cjs` example, we will see the callback style version of the `after` method:

```
const Fastify = require('fastify')
const fp = require('fastify-plugin')

async function boot () {
  // [1]
  async function plugin1 (fastify, options) {
    fastify.decorate('plugin1Decorator', 'plugin1')
  }
  async function plugin2 (fastify, options) {
    fastify.decorate('plugin2Decorator', 'plugin2')
  }
  async function plugin3 (fastify, options) {
    fastify.decorate('plugin3Decorator', 'plugin3')
```

```
  }

  const app = Fastify({ logger: true })

  await app // [2]
    .register(fp(plugin1))
    .register(fp(plugin2))
    .after((_error) => { // [3]
      console.log('plugin1Decorator',
        app.hasDecorator('plugin1Decorator'))
      console.log('plugin2Decorator',
        app.hasDecorator('plugin2Decorator'))
    })
    .register(fp(plugin3)) // [4]

  console.log('plugin3Decorator',
    app.hasDecorator('plugin3Decorator'))

  await app.ready()
  console.log('app ready')
}
boot()
```

We declare and then register the same three plugins from the last examples ([**1**]). On [**2**], we start our chain of method calls, adding the `after` call ([**3**]) before the last `register` ([**4**]). Inside the `after` callback, we are sure, if the error is null, that the first two plugins are loaded correctly; in fact, the decorators have the correct values. If we run this snippet, we will have the same result as the previous one.

Fastify guarantees that all `after` callbacks will be called before the application's ready event is dispatched. This method can be helpful if we prefer chaining instance methods but still need control over the boot sequence.

A declaration order

Even if Fastify keeps the correct order of plugin registration, we should still follow some good practices to maintain more consistent and predictable boot behavior. If something terrible happens during the loading, using the following declaration order will help figure out what the issue is:

- Plugins installed from npmjs.com
- Our plugins
- Decorators

- Hooks

- Services, routes, and so on

This declaration order guarantees that all stuff declared in the current context will be accessible. Since in Fastify, everything can, and really should, be defined inside a plugin, we should replicate the preceding structure at every registration level.

The boot sequence is the series of operations that Fastify performs to load all of the plugins that are registered and start the server. We learned how this process is deterministic, since the order of registration matters. Finally, we covered how developers have two methods at their disposal to fine-tune the boot sequence, despite Fastify providing a default behavior.

During the boot sequence, one or more errors can occur. In the next section, we will cover the most common errors and learn how to deal with them.

Handling boot and plugin errors

This section will analyze some of the most common Fastify boot errors and how we can deal with them. But what are boot errors anyway? All errors thrown during the initial load of our application, before the ready event is dispatched and the server listens for incoming connections, are called **boot errors**.

These errors are usually thrown when something unexpected happens during the registration of a plugin. If there are any unhandled errors during a plugin registration, Fastify, with the help of `Avvio`, will notify us and stop the boot process.

We can distinguish between two error types:

- Errors that we can recover from

- Errors that are not recoverable by any means

First, we will look at the most common non-recoverable error, `ERR_AVVIO_PLUGIN_TIMEOUT`, which usually means we forgot to tell Fastify to continue with the boot process.

Then, we will learn about the tools Fastify gives us to recover from other kinds of errors. It is important to note that we don't want to recover from an error more often than not, and it is better to just make the server crash during the boot!

ERR_AVVIO_PLUGIN_TIMEOUT

To prevent the boot process from being stuck indefinitely, Fastify has set a maximum amount of time a plugin can take to load. If a plugin takes longer than that, Fastify will throw an `ERR_AVVIO_PLUGIN_TIMEOUT` error. The default timeout is set to 10 seconds, but the value can be easily changed using the `pluginTimeout` server option.

This is one of the most common boot errors, and usually, it happens for two reasons:

- We forgot to call the done callback when registering a plugin

- The registration promise isn't resolved in time

It is also the most confusing one, and it is certainly a non-recoverable error. The code written in timeout-error.cjs generates this error on purpose, allowing us to analyze the stack trace to understand how we can spot it and find where it originates:

```
const Fastify = require('fastify')

const app = Fastify({ logger: true })

app.register(function myPlugin(fastify) {
  console.log('Registering my first plugin.')
})

app.ready()
  .then(() => {
    console.log('app ready')
})
```

Here, we register our plugin. We are giving it a name because, as we will see, it will help with the stack trace. Even if the plugin is not an async function, we deliberately don't call the done callback to let Fastify know that the registration went without errors.

If we run this snippet, we will receive the ERR_AVVIO_PLUGIN_TIMEOUT error:

```
$ node timeout-error.cjs
Registering my first plugin.
Error: ERR_AVVIO_PLUGIN_TIMEOUT: plugin did not start in time:
myPlugin. You may have forgotten to call 'done' function or to resolve
a Promise
    at Timeout._onTimeout (/node_modules/avvio/plugin.js:123:19)
    at listOnTimeout (internal/timers.js:554:17)
    at processTimers (internal/timers.js:497:7) {
  code: 'ERR_AVVIO_PLUGIN_TIMEOUT',
  fn: [Function: myPlugin]
}
```

As we can see immediately, the error is pretty straightforward. Fastify uses the function name, myPlugin, to point us in the correct direction. Moreover, it suggests that we might forget to call 'done' or resolve the promise. It is worth mentioning that, in real-world scenarios, this error is usually thrown when there are some connection-related problems – for example, the database is not reachable at the time of the plugin registration.

Recovery from a boot error

Usually, if an error happens during boot time, something serious prevents our application from starting. In such cases, as we already saw, the best thing Fastify can do is to stop the boot process and notify us about the errors it encountered. However, there are a few cases where we can recover from an error and continue with the boot process. For example, we have an optional plugin to measure application metrics, and we want to start the application even if it is not loading correctly.

`error-after.cjs` shows how to recover from the error using our old friend, the `after` method:

```
const Fastify = require('fastify')

// [1]
async function plugin1(fastify, options) {
  throw new Error('Kaboom!')
}

const app = Fastify({ logger: true })

app
  .register(plugin1)
  .after(err => {
    if (err) { // [2]
      console.log(
        `There was an error loading plugin1:
          '${err.message}'. Skipping.`
      )
    }
  })

app.ready()
  .then(() => {
    console.log('app ready')
})
```

First, we declare a dummy plugin ([1]) that always throws. Then, we register it and use the `after` method to check for registration errors ([2]). If any error is found, we log it to the console. Calling `after` has the effect of "catching" the boot error, and therefore, the boot process will not stop since we handled the unexpected behavior. We can run the snippet to check that it works as expected:

```
$ node error-after.cjs
There was an error loading plugin1: 'Kaboom!'. Skipping.
app ready
```

Since the last logged line is `app ready`, we know that the boot process went well, and our application has started!

Summary

In this chapter, we learned about the importance of plugins and how the Fastify boot process works. Everything in Fastify can and really should be put in a plugin. It is the base building block of scalable and maintainable applications, thanks to encapsulation and a predictable loading order. Furthermore, we can use `fastify-plugin` to control the default encapsulation and manage dependencies between plugins.

We learned how our applications are nothing more than a bunch of Fastify plugins that work together. Some are used to encapsulate routers, using prefixes to namespace them. Others are used to add core functionalities, such as connections to databases or some other external systems. In addition to that, we can install and use core and community plugins directly from npm.

Moreover, we covered the boot process's asynchronous nature and how every step can be awaited if needed. It is guaranteed that if any error is encountered during the loading of plugins, the boot process will stop, and the error will be logged to the console.

In the next chapter, we will learn all the ways to declare the endpoints of our application. We will see how to add route handlers and how to avoid the major pitfalls. Finally, what we learned in this chapter about plugins will be helpful when dealing with route scoping and route grouping!

3

Working with Routes

Applications would not exist without routes. Routes are the doors to those clients that need to consume your API. In this chapter, we will present more examples, focusing on how to become more proficient at managing routes and keeping track of all our endpoints. We will cover all the possibilities that Fastify offers to define new endpoints and manage the application without giving you a headache.

It is worth mentioning that Fastify supports async/await handlers out of the box, and it is crucial to understand its implications. You will look at the difference between sync and async handlers, and you will learn how to avoid the major pitfalls. Furthermore, we will learn how to handle URL parameters, the HTTP request's body input, and query strings too.

Finally, you will understand how the router works in Fastify, and you will be able to control the routing to your application's endpoint as never before.

In this chapter, we will cover the following topics:

- Declaring API endpoints and managing errors
- Routing to the endpoint
- Reading the client's input
- Managing the route's scope
- Adding new behaviors to routes

Technical requirements

As mentioned in the previous chapters, you will need the following:

- A working Node.js 18 installation
- A text editor to try the example code
- An HTTP client to test out code such as CURL or Postman

All the snippets in this chapter are available on GitHub at `https://github.com/PacktPublishing/Accelerating-Server-Side-Development-with-Fastify/tree/main/Chapter%203`.

Declaring API endpoints and managing the errors

An endpoint is an interface that your server exposes to the outside, and every client with the coordinates of the route can call it to execute the application's business logic.

Fastify lets you use the software architecture you like most. In fact, this framework doesn't limit you from adopting **Representation State Transfer** (**REST**), **GraphQL**, or simple **Application Programming Interfaces** (**APIs**). The first two architectures standardize the following:

- The application endpoints: The standard shows you how to expose your business logic by defining a set of routes

- The server communication: This provides insights into how you should define the input/output

In this chapter, we will create simple **API** endpoints with JSON input/output interfaces. This means that we have the freedom to define an internal standard for our application; this choice will let us focus on using the Fastify framework instead of following the standard architecture.

In any case, in *Chapter 7*, we will learn how to build a **REST** application, and in *Chapter 14*, we will find out more about using **GraphQL** with Fastify.

> Too many standards
>
> Note that the **JSON:API** standard exists too: `https://jsonapi.org/`. Additionally, Fastify lets you adopt this architecture, but that topic will not be discussed in this book. Check out `https://backend.cafe/` to find more content about Fastify and this book!

In the following sections, we assume you already understand the anatomy of an HTTP request and the differences between its parts, such as a query parameter versus body input. A great resource to rehearse these concepts is the Mozilla site: `https://developer.mozilla.org/en-US/docs/Web/HTTP/Messages`.

Declaration variants

In previous chapters, we learned how to create a Fastify server, so we will assume you can create a new project (or, if you have trouble doing so, you can read *Chapter 1*).

In the same chapter, we pointed out the two syntaxes you can use to define routes:

- The generic declaration, using `app.route(routeOptions)`

- The shorthand syntax: `app.<HTTP method>(url[, routeOptions], handler)`

The second statement is more expressive and readable when you need to create a small set of endpoints, whereas the first one is extremely useful for adding automation and defining very similar routes. Both declarations expose the same parameters, but it is not only a matter of preference. Having to choose one over the other can negatively impact your code base at scale. In this chapter, we will learn how to avoid this pitfall and how to choose the best syntax based on your needs.

Before starting our coding, we will get a quick overview of the `routeOptions` object, which we will use in the next sections to develop our baseline knowledge, which you can refer to in your future projects.

The route options

Before learning how to code the application's routes further, we must preview the `routeOptions` properties (note that some of them will be discussed in the following chapters).

The options are listed as follows:

- `method`: This is the HTTP method to expose.
- `url`: This is the endpoint that listens for incoming requests.
- `handler`: This is the route business logic. We met this property in previous chapters.
- `logLevel`: This is a specific log level for a single route. We will find out how useful this property can be in *Chapter 11*.
- `logSerializer`: This lets you customize the logging output of a single route, in conjunction with the previous option.
- `bodyLimit`: This limits the request payload to avoid possible misuse of your endpoints. It must be an integer that represents the maximum number of bytes accepted, overwriting the root instance settings.
- `constraints`: This option improves the routing of the endpoint. We will learn more about how to use this option in the *Routing to the endpoint* section.
- `errorHandler`: This property accepts a special handler function to customize how to manage errors for a single route. The following section will show you this configuration.
- `config`: This property lets you specialize the endpoint by adding new behaviors.
- `prefixTrailingSlash`: This option manages some special usages during the route registration with plugins. We will talk about that in the *Routing to the endpoint* section.
- `exposeHeadRoute`: This Boolean adds or removes a HEAD route whenever you register a GET one. By default, it is `true`.

Then, there are many highly specialized options to manage the request validation: `schema`, `attachValidation`, `validatorCompiler`, `serializerCompiler`, and `schemaErrorFormatter`. All these settings will be covered in *Chapter 5*.

Finally, you must be aware that every route can have additional hooks that will only run for the route itself. You can just use the hooks names info and the `routeOptions` object to attach them. We will see an example at the end of this chapter. The hooks are the same as we listed in *Chapter 1*: `onRequest`, `preParsing`, `preValidation`, `preHandler`, `preSerialization`, `onSend`, and `onResponse`, and they will take action during the **request life cycle**.

It is time to see these options in action, so let's start defining some endpoints!

Bulk routes loading

The generic declaration lets you take advantage of the route automation definition. This technique aims to divide the source code into small pieces, making them more manageable while the application grows.

Let's start by understanding the power of this feature:

```
const routes = [
  {
    method: 'POST', url: '/cat',
    handler: function cat (request, reply) {
      reply.send('cat') }
  },
  {
    method: 'GET', url: '/dog',
    handler: function dog (request, reply) {
      reply.send('dog') }
  }
]
routes.forEach((routeOptions) => {
  app.route(routeOptions)
})
```

We have defined a `routes` array where each element is Fastify's `routeOptions` object. By iterating the `routes` variable, we can add the routes programmatically. This will be useful if we split the array by context into `cat.cjs` and `dog.cjs`. Here, you can see the `cat.cjs` code example:

```
module.exports = [
  {
    method: 'POST',
    url: '/cat',
    handler: function cat (request, reply) {
      reply.send('cat') }
```

```
    }
]
```

By doing the same for the /dog endpoint configuration, the server setup can be changed to the following:

```
const catRoutes = require('./cat.cjs')
const dogRoutes = require('./dog.cjs')

catRoutes.forEach(loadRoute)
dogRoutes.forEach(loadRoute)

function loadRoute(routeOptions) {
  app.route(routeOptions)
}
```

As you can see, the route loading seems more precise and straightforward. Moreover, this code organization gives us the ability to easily split the code and let each context grow at its own pace, reducing the risk of creating huge files that could be hard to read and maintain.

We have seen how the generic app.route() method can set up an application with many routes, centralizing the loading within the server definition and moving the endpoint logic to a dedicated file to improve the project's legibility.

Another way to improve the code base is by using async/await in the route handler, which Fastify supports out of the box. Let's discuss this next.

Synchronous and asynchronous handlers

In *Chapter 1*, we saw an overview of the essential role of the route handler and how it manages the Reply component.

To recap briefly, there are two main syntaxes we can adopt. The sync syntax uses callbacks to manage asynchronous code and it must call reply.send() to fulfill the HTTP request:

```
function syncHandler (request, reply) {
  readDb(function callbackOne (err, data1) {
    readDb(function callbackTwo (err, data2) {
      reply.send(`read from db ${data1} and ${data2}`)
    })
  })
}
```

In this example, we are simulating two async calls to a `readDb` function. As you can imagine, adding more and more asynchronous I/O such as files read or database accesses could make the source quickly unreadable, with the danger of falling into callback hell (you can read more about this at `http://callbackhell.com/`).

You can rewrite the previous example, using the second syntax to define a route handler, with an async function:

```
async function asyncHandler (request, reply) {
  const data1 = await readDb()
  const data2 = await readDb()
  return `read from db ${data1} and ${data2}`
}
```

As you can see, regardless of how many async tasks your endpoint needs to run, the function body can be read sequentially, making it much more readable. This is not the only syntax an async handler supports, and there are other edge cases you could encounter.

Reply is a Promise

In an `async` function handler, it is highly discouraged to call `reply.send()` to send a response back to the client. Fastify knows that you could find yourself in this situation due to a legacy code update. If this happens, the solution is to return a `reply` object. Here is a quick real-(bad) world scenario:

```
async function newEndpoint (request, reply) {
  if (request.body.type === 'old') { // [1]
    oldEndpoint(request, reply)
    return reply // [2]
  } else {
    const newData = await something(request.body)
    return { done: newData }
  }
}
```

In this example endpoint, the `if` statement of **[1]** runs the `oldEndpoint` business logic that manages the `reply` object in a different way compared to the `else` case. In fact, the `oldEndpoint` handler was implemented in the callback style. So, how do we tell Fastify that the HTTP response has been delegated to another function? You just need to return the `reply` object of **[2]**! The `Reply` component is a `thenable` interface. This means that it implements the `.then()` interface in the same way as the `Promise` object! Returning it is like producing a promise that will be fulfilled only when the `.send()` method has been executed.

The readability and flexibility of the async handlers are not the only advantages: what about errors? Errors can happen during the application runtime, and Fastify helps us deal with them with widely used defaults.

How to reply with errors

Generally, in Fastify, an error can be **sent** when the handler function is sync or **thrown** when the handler is async. Let's put this into practice:

```
function syncHandler (request, reply) {
  const myErr = new Error('this is a 500 error')
  reply.send(myErr) // [1]
}
async function ayncHandler (request, reply) {
  const myErr = new Error('this is a 500 error')
  throw myErr // [2]
}
async function ayncHandlerCatched (request, reply) {
  try {
    await ayncHandler(request, reply)
  } catch(err) { // [3]
    this.log.error(err)
    reply.status(200)
    return { success: false }
  }
}
```

As you can see, at first sight, the differences are minimal: in [1], the send method accepts a Node.js Error object with a custom message. The [2] example is quite similar, but we are throwing the error. The [3] example shows how you can manage your errors with a try/catch block and choose to reply with a 200 HTTP success in any case!

Now, if we try to add the error management to the syncHandler example, as shown previously, the sync function becomes the following:

```
function syncHandler (request, reply) {
  readDb(function callbackOne (err, data1) {
    if (err) {
      reply.send(err)
      return
    }
    readDb(function callbackTwo (err, data2) {
      if (err) {
        reply.send(err)
        return
      }
      reply.send(`read from db ${data1} and ${data2}`)
    })
  })
}
```

The `callback` style strives to be lengthy and hard to read. Instead, the `asyncHandler` function shown in the code block of the *Synchronous and asynchronous* section doesn't need any updates. This is because the error thrown will be managed by Fastify, which will send the error response to the client.

So far, we have seen how to reply to an HTTP request with a Node.js `Error` object. This action sends back a JSON payload with a 500 status code response if you didn't set it using the `reply.code()` method that we saw in *Chapter 1*.

The default JSON output is like the following:

```
{
  "statusCode": 500,
  "error": "Internal Server Error",
  "message": "app error"
}
```

The `new Error('app error')` code creates the error object that produces the previous output message.

Fastify has many ways to customize the error response, and usually, it depends on how much of a hurry you are in. The options are listed as follows:

- Adopt the default Fastify output format: This solution is ready to use and optimized to speed up the error payload serialization. It works great for rapid prototyping.

- Customize the error handler: This feature gives you total control of the application errors.

- Custom response management: This case includes a call to `reply.code(500).send(myJsonError)` providing a JSON output.

Now we can better explore these options.

Adopting the default Fastify error output is very simple because you need to **throw** or **send** an `Error` object. To tweak the body response, you can customize some `Error` object fields:

```
const err = new Error('app error') // [1]
err.code = <ERR-001> // [2]
err.statusCode = 400 // [3]
```

This example configuration maps the following:

1. The String message, which is provided in the `Error` constructor as the `message` field.

2. The optional `code` field to the same JSON output key.

3. The `statusCode` parameter, which will change the HTTP status code response and the `error` string. The output string is set by the default Node.js `http.STATUS_CODES` module.

The result of the previous example will produce the following output:

```
{
  "statusCode": 400,
  "code": "ERR001",
  "error": "Bad Request",
  "message": "app error"
}
```

This payload might not be informative for the client because it contains a single error. So, if we want to change the output to an array of errors when more than one error happens, such as form validation, or if you need to change the output format to adapt it to your API ecosystem, you must know the ErrorHandler component:

```
app.setErrorHandler(function customErrorHandler(error,
request, reply) {
  request.log.error(error)
  reply.send({ ops: error.message })
})
```

The error handler is a function that is executed whenever an Error object or a JSON is **thrown** or **sent**; this means that the error handler is the same regardless of the implementation of the route. Earlier, we said that a JSON is **thrown**: trust me, and we will explain what that means later in this section.

The error handler interface has three parameters:

- The first one is the object that has been thrown or the Error object that has been sent.
- The second is the Request component that originated the issue
- The third is the Reply object to fulfill the HTTP request as a standard route handler

The error handler function might be an async function or a simple one. As for the route handler, you should return the response payload in case of an async function, and call reply.send() for the sync implementation. In this context, you can't throw or send an Error instance object. This would create an infinite loop that Fastify manages. In this case, it will skip your custom error handler, and it will call the parent scope's error handler or the default one if it is not set. Here is a quick example:

```
app.register(async function plugin (pluginInstance) {
  pluginInstance.setErrorHandler(function first (error,
  request, reply) {
    request.log.error(error, 'an error happened')
    reply.status(503).send({ ok: false }) // [4]
  })

  pluginInstance.register(async function childPlugin (deep,
  opts) {
```

```
    deep.setErrorHandler(async function second (error,
    request, reply) {
      const canIManageThisError = error.code === 'yes, you
      can' // [2]
      if (canIManageThisError) {
        reply.code(503)
        return { deal: true }
      }
      throw error // [3]
    })

    // This route run the deep's error handler
    deep.get('/deepError',
      () => { throw new Error('ops') }) // [1]   })
  })
```

In the preceding code snippet, we have a `plugin` function that has a `childPlugin` context. Both these encapsulated contexts have one custom error handler function. If you try to request GET / deep [1], an error will be thrown. It will be managed by the `second` error handler function that will decide whether to handle it or re-throw it [2]. When the failure is re-thrown [3], the parent scope will intercept the error and handle it [4]. As you can see, you can implement a series of functions that will handle a subset of the application's errors.

It is crucial to keep in mind that you should take care of the response status code when you implement your error handler; otherwise, it will be **500 – Server Error** by default.

As we saw in the preceding example, the error handler can be assigned to the application instance and a plugin instance. This will set up the handler for all the routes in their context. This means that the error handler is fully encapsulated, as we learned in *Chapter 2*.

Let's see a quick example:

```
async function errorTrigger (request, reply) {
  throw new Error('ops')
}
app.register(async function plugin (pluginInstance) {
  pluginInstance.setErrorHandler(function (error, request,
  reply) {
    request.log.error(error, 'an error happened')
    reply.status(503).send({ ok: false })
  })
  pluginInstance.get('/customError', errorTrigger) // [1]
})
app.get('/defaultError', errorTrigger) // [2]
```

We have defined a bad route handle, `errorTrigger`, that will always throw an `Error`. Then, we registered two routes:

- The GET `/customError` [1] route is inside a plugin, so it is in a new Fastify context.
- The root application instance registers the GET `/defaultError` [2] route instead.

We set `pluginInstance.setErrorHandler`, so all the routes registered inside that plugin and its children's contexts will use your custom error handler function during the plugin creation. Meanwhile, the app's routes will use the default error handler because we didn't customize it.

At this stage, making an HTTP request to those endpoints will give us two different outputs, as expected:

- The GET `/customError` route triggers the error, and it is managed by the custom error handler, so the output will be `{ "ok": false }`.
- The GET `/defaultError` endpoint replies with the Fastify default JSON format that was shown at the beginning of this section.

It is not over yet! Fastify implements an outstanding **granularity** for most of the features it supports. This means that you can set a custom error handler for every route!

Let's add a new endpoint to the previous example:

```
app.get('/routeError', {
  handler: errorTrigger,
  errorHandler: async function (error, request, reply) {
    request.log.error(error, 'a route error happened')
    return { routeFail: false }
  }
})
```

First of all, during the endpoint definition, we must provide the `routeOptions` object to set the custom `errorHandler` property. The function parameter is the same as the `setErrorHandler` method. In this case, we switched to an async function: as already mentioned, this format is supported too.

Finally, the last option you might implement to return an error is calling `reply.send()`, like you would do when sending data back to the client:

```
app.get('/manualError', (request, reply) => {
  try {
    const foo = request.param.id.split('-') // this line
                                            //       throws
    reply.send('I split the id!')
  } catch (error) {
    reply.code(500).send({ error: 'I did not split the id!' })
  }
})
```

This is a trivial solution, but you must keep in mind the possibilities Fastify offers you. It is essential to understand that this solution is a managed error, so Fastify is unaware that an error has been caught in your handler. In this case, you must set the HTTP response status code; otherwise, it will be **200 – Success** by default. This is because the route has been executed successfully from the framework's point of view. In this case, the `errorHandler` property is not performed either. This could impact your application logging and system monitoring, or limit the code you will be able to reuse in your code base, so use it consciously.

In the previous code, we called the `split` function in an `undefined` variable in a synchronous function. This will trigger a `TypeError`. If we omit the `try/catch` block, Fastify will handle the error, preventing the server crashing.

Instead, if we moved this implementation error to a callback, the following will be the result:

```
app.get('/fatalError', (request, reply) => {
  setTimeout(() => {
    const foo = request.param.id.split('-')
    reply.send('I split the id!')
  })
})
```

This will break our server because there is an uncaught exception that no framework can handle. This issue that you might face is typically related to sync handlers.

In this section, we covered many topics; we learned more about route declaration and the differences between sync and async handlers, and how they reply to the clients with an error.

The takeaways can be outlined as follows:

	Async handler	**Sync handler**
Interface	`async function handler(request, reply) {}`.	`Function handler(request, reply) {}`.
How to reply	Return the payload.	Call `reply.send()`.
Special usage for reply	If you call `reply.send()`, you must return the `reply` object.	You can return the reply object safely.
How to reply with an Error	Throw an error.	Call `reply.send(errorInstance)`.
Special usage for errors	None.	You can throw only in the main scope's handler function.

Figure 3.1 – The differences between the async and sync handlers

Now you have a solid understanding of how to define your endpoints and how to implement the handler functions. You should understand the common pitfalls you might face when writing your business logic, and you have built up the critical sense to choose the best route definition syntax based on your needs.

We are ready to take a break from route handling and move on to an advanced topic: **routing**.

Routing to the endpoint

Routing is the phase where Fastify receives an HTTP request, and it must decide which handler function should fulfill that request. That's it! It seems simple, but even this phase is optimized to be performant and to speed up your server.

Fastify uses an external module to implement the router logic called **find-my-way** (https://github.com/delvedor/find-my-way). All its features are exposed through the Fastify interface so that Fastify will benefit from every upgrade to this module. The strength of this router is the algorithm implemented to find the correct handler to execute.

> **The router under the hood**
>
> You might find it interesting that find-my-way implements the **Radix-tree** data structure, starting from the route's URLs. The router traverses the tree to find the matched string URL whenever the server receives an HTTP request. Every route, tracked into the tree, carries all the information about the Fastify instance it belongs to.

During the startup of the **application instance**, every route is added to the router. For this reason, Fastify doesn't throw errors if you write the following:

```
app.get('/', async () => {})
app.get('/', async () => {}) // [1]
```

When you write the same route for the second time, **[1]**, you might expect an error. This will only happen when you execute one of the ready, listen, or inject methods:

```
app.listen(8080, (err) => {
  app.log.error(err)
})
```

The double route registration will block the server from starting:

```
$ node server.mjs
AssertionError [ERR_ASSERTION]: Method 'GET' already declared for
route '/' with constraints '{}'
```

The preceding example shows you the asynchronous nature of Fastify and, in this specific case, how the routes are loaded.

The router has built the Radix-tree carrying the route handler and Fastify's context. Then, Fastify relies on the context's immutability. This is why it is not possible to add new routes when the server has started. This might seem like a limitation, but by the end of this chapter, you will see that it is not.

We have seen how the router loads the URL and lookups for the handler to execute, but what happens if it doesn't find the HTTP request URL?

The 404 handler

Fastify provides a way to configure a 404 handler. It is like a typical route handler, and it exposes the same interfaces and async or sync logic:

```
app.setNotFoundHandler(function custom404 (request, reply) {
  const payload = {
    message: 'URL not found'
  }
  reply.send(payload)
})
```

Here, we have registered a new handler that always returns the same payload.

By default, the 404 handler replies to the client with the same format as the default error handler:

```
{
  "message": "Route GET:/ops not found",
  "error": "Not Found",
  "statusCode": 404
}
```

This JSON output keeps the consistency between the two events: the error and the route not found, replying to the client with the same response fields.

As usual, this feature is also encapsulated, so you could set one Not Found handler for each context:

```
app.register(async function plugin (instance, opts) {
  instance.setNotFoundHandler(function html404 (request,
  reply) {
    reply.type('application/html').send(niceHtmlPage)
  })
}, { prefix: '/site' }) // [1]

app.setNotFoundHandler(function custom404 (request, reply)
{
  reply.send({ not: 'found' })
})
```

In this example, we have set the `custom404` root 404 handler and the plugin instance `html404`. This can be useful when your server manages multiple contents, such as a static website that shows a cute and funny HTML page when an non-existent page is requested, or shows a JSON when a missing API is queried.

The previous code example tells Fastify to search for the handler to execute into the plugin instance when the requested URL starts with the `/site` string. If Fastify doesn't find a match in this context, it will use the Not Found handler of that context. So, for example, let's consider the following URLs:

- The `http://localhost:8080/site/foo` URL will be served by the `html404` handler
- The `http://localhost:8080/foo` URL will be served by the `custom404` instead

The `prefix` parameter (marked as **[1]** in the previous code block) is mandatory to set multiple 404 handlers; otherwise, Fastify will not start the server, because it doesn't know when to execute which one, and it will trigger a startup error:

```
Error: Not found handler already set for Fastify instance with prefix:
'/'
```

Another important aspect of the Not Found handling is that it triggers the **request life cycle hooks** registered in the context it belongs to. We got a quick introduction to hooks in *Chapter 1*, and we will further explain this Fastify feature in *Chapter 4*.

Here, the takeaway is how do you know if the hook has been triggered by an HTTP request with or without a route handler? The answer is the `is404` flag, which you can check as follows:

```
app.addHook('onRequest', function hook (request, reply,
done) {
  request.log.info('Is a 404 HTTP request? %s',
  request.is404)
  done()
})
```

The `Request` component knows whether the HTTP request is fulfilled by a route handler or by a Not Found one, so you can skip some unnecessary request process into your hook functions, filtering those requests that will not be handled.

So far, you have learned how to manage the response when an URL doesn't match any of your application endpoints. But what happens if a client hits a 404 handler, due to a wrong trailing slash?

Router application tuning

Fastify is highly customizable in every component: the router is one of them. You are going to learn how to tweak the router settings, to make the router more flexible and deal with a client's common trouble. It is important to understand these settings to anticipate common issues and to build a great set of APIs on the first try!

The trailing slash

The trailing slash is the / character when it is the last character of the URL, query parameter excluded.

Fastify thinks that the /foo and /foo/ URLs are different, and you can register them and let them reply to two completely different outputs:

```
app.get('/foo', function (request, reply) {
  reply.send('plain foo')
})
app.get('/foo/', function (request, reply) {
  reply.send('foo with trailin slash')
})
```

Often, this interface can be misunderstood by clients. So, you can configure Fastify to treat those URLs as the same entity:

```
const app = fastify({
  ignoreTrailingSlash: true
})
```

The ignoreTrailingSlash setting forces the Fastify router to ignore the trailing slash for **all the application's routes**. Because of this, you won't be able to register the /foo and /foo/ URLs, and you will receive a startup error. Doing so will consume your API, but you will not have to struggle with the 404 errors if the URL has been misprinted with an ending / character.

Case-insensitive URLs

Another common issue you could face is having to support both the /fooBar and /foobar URLs as a single endpoint (note the case of the B character). As per the trailing slash example, Fastify will manage these routes as two distinct items; in fact, you can register both routes with two different handler functions, unless you set the code in the following way:

```
const app = fastify({
  caseSensitive: false
})
```

The `caseSensitive` option will instruct the router to match all your endpoints in lowercase:

```
app.get('/FOOBAR', function (request, reply) {
  reply.send({
    url: request.url, // [1]
    routeUrl: request.routerPath // [2]
  })
})
```

The `/FOOBAR` endpoint will reply to all possible combinations, such as `/FooBar`, `/foobar`, `/fooBar`, and more. The handler output will contain both the HTTP request URL, [1], and the route one, [2]. These two fields will match the setup without changing them to lowercase.

So, for example, making an HTTP request to the `GET   /FoObAr` endpoint will produce the following output:

```
{
  "url": "/FoObAr",
  "routeUrl": "/FOOBAR"
}
```

Using the case-insensitive URL, matching the setup could look odd. In fact, it is highly discouraged to do so, but we all know that legacy code exists, and every developer must deal with it. Now, if you have to migrate many old endpoints implemented using a case-insensitive router, you know how to do it.

> **The URL's pathname**
> If you are struggling while choosing whether your endpoint should be named `/fast-car` or `/fast_car`, you should know that a hyphen is broadly used for web page URLs, whereas the underscore is used for API endpoints.

Another situation you might face during a migration is having to support old routes that will be discarded in the future.

Rewrite URL

This feature adds the possibility of changing the HTTP's requested URL before the routing takes place:

```
const app = fastify({
  rewriteUrl: function rewriteUrl (rawRequest) {
    if (rawRequest.url.startsWith('/api')) {
      return rawRequest.url
    }
    return `/public/${rawRequest.url}`
  }
})
```

The `rewriteUrl` parameter accepts an input sync function that must return a string. The returned line will be set as the request URL, and it will be used during the routing process. Note that the function argument is the standard `http.IncomingMessage` class and not the Fastify `Request` component.

This technique could be useful as a URL expander or to avoid redirecting the client with the 302 HTTP response status code.

> **Logging the URL rewrite**
>
> Unfortunately, the `rewriteUrl` function will not be bound to the Fastify root instance. This means you will not be able to use the `this` keyword in that context. Fastify will log the debug information if the function returns a different URL than the original. In any case, you will be able to use the `app.log` object at your convenience.

We have explored how to make Fastify's router more flexible in order to support a broad set of use cases that you may encounter in your daily job.

Now, we will learn how to configure the router to be even more granular.

Registering the same URLs

As we have seen previously, Fastify doesn't register the same HTTP route path more than once. This is a limit, due to the Fastify router. The router searches the correct handler to execute by matching with the following rules:

- The request HTTP method
- The request string URL

The search function must only return one handler; otherwise, Fastify can't choose which one to execute. To overcome the limit, with Fastify, you can extend these two parameters to all the request's parts, such as headers and request metadata!

This feature is the **route constraint**. A constraint is a check performed when the HTTP request has been received by the server and must be routed to an endpoint. This step reads the raw `http.IncomingMessage` to pull out the value to apply the constraint check. Essentially, you can see two main logic steps:

1. The constraint configuration in the route option means that the endpoint can only be reached if the HTTP request meets the condition.
2. The constraint evaluation happens when the HTTP request is routed to a handler.

A constraint can be mandatory if derived from the request, but we will look at an example later.

Now, let's assume we have an endpoint that must change the response payload. This action would be a breaking change for our customers. A breaking change means that all the clients connected to our application must update their code to read the data correctly.

In this case, we can use **route versioning**. This feature lets us define the same HTTP route path with a different implementation, based on the version requested by the client. Consider the following working snippet:

```
app.get('/user', function (request, reply) {
  reply.send({ user: 'John Doe' })
})
app.get('/user',
  {
    constraints: {
      version: '2.0.0'
    }
  },
  function handler (request, reply) {
    reply.send({ username: 'John Doe' })
  }
)
```

We have registered the same URL with the same HTTP method. The two routes reply with a different response object that is not backward compatible.

> **Backward compatibility**
>
> An endpoint is backward compatible when changes to its business logic do not require client updates.

The other difference is that the second endpoint has a new `constraints` option key, pointing to a JavaScript object input. This means that the router must match the URL path, the HTTP method, and all the constraints to apply that handler.

By default, Fastify supports two constraints:

- The `host` constraint checks the `host` request header. This check is not mandatory, so if the request has the `host` header, but it doesn't match with any constrained route, a generic endpoint without a constraint can be selected during the routing.

- The `version` constraint analyzes the `accept-version` request header. When a request contains this header, the check is mandatory, and a generic endpoint without a constraint can't be considered during the routing.

To explain these options better, let's see them in action:

```
app.get('/host', func0)
app.get('/host', {
  handler: func2,
  constraints: {
    host: /^bar.*/
  }
})
app.get('/together, func0)
app.get('/together', {
  handler: func1,
  constraints: {
    version: '1.0.1',
    host: 'foo.fastify.dev'
  }
})
```

The /host handler only executes when a request has the host header that starts with bar, so the following command will reply with the func2 response:

```
$ curl --location --request GET "http://localhost:8080/host" --header
"host: bar.fastify.dev"
```

Instead, setting the host header as foo.fastify.dev will execute the func0 handler; this happens because the host constraint is not mandatory, and an HTTP request with a value can match a route that has no constraint configured.

The /together endpoint configures two constraints. The handler will only be executed if the HTTP request's header has both the corresponding HTTP headers:

```
$ curl --location --request GET "http://localhost:8080/together"
--header "accept-version: 1.x" --header "host: foo.fastify.dev"
```

The host match is a simple string match; instead, the accept-version header is a **Semantic Versioning (SemVer)** range string matching.

The SemVer is a specification to name a release of software in the Node.js ecosystem. Thanks to its clarity, it is broadly used in many contexts. This naming method defines three numbers referred to as **major.minor.patch**, such as 1.0.1. Each number indicates the type of software changes that have been published:

- Major version: The change is not backward compatible, and the client must update how it processes the HTTP request.

- Minor bump version: A new feature is added to the software, keeping the endpoint I/O backward compatible.

- Patch version: Bug fixes that improve the endpoint without changing the exposed API.

The specification defines how to query a SemVer string version, too. Our use case focuses on the `1.x` range, which means *the most recent major version 1*; the 1.0.x translates to *the most recent major version 1, and minor equals 0*. For a complete overview of the SemVer query syntax, you can refer to `https://semver.org/`.

So, the `version` constraint supports the SemVer query syntax as an HTTP header value to match the target endpoint.

Note that, in this case, when a request has the `accept-version` header, the check is mandatory. This means that routes without a constraint can't be used. Here, the rationale is that if the client wants a well-defined route version, it cannot reply from an unversioned route.

Multiple constraint match

Note that the constraints can face conflicts during the evaluation. If you define two routes with the same host regular expression, an HTTP request might match both of them. In this case, the last registered route will be executed by the router. It would be best if you avoided these cases by configuring your constraints carefully.

As mentioned already, you can have many more constraints to route the HTTP request to your handlers. In fact, you can add as many constraints as you need to your routes, but it will have a performance cost. The routing selects the routes that match the HTTP method and path and will then process the constraints for every incoming request. Fastify gives you the option to implement custom constraints based on your needs. Creating a new constraint is not the goal of this book, but you can refer to this module at `https://github.com/Eomm/header-constraint-strategy`, which is maintained by the co-author of this book. Your journey with Fastify is not restricted to this practical book!

At this stage, we have understood how to add a route and how to drive the HTTP requests to it. Now we are ready to jump into input management.

Reading the client's input

Every API must read the client's input to behave. We already mentioned the four HTTP request input types, which are supported by Fastify:

- The path parameters are positional data, based on the endpoint's URL format
- The query string is an additional part of the URL the client adds to provide variable data
- The headers are additional `key:value` pairs that pair information passing between the client and the server
- The body is the request payload that contains the client's data submission

Let's take a look at each in more detail.

The path parameters

The path parameters are variable pieces of a URL that could identify a resource in our application server. We already covered this aspect in *Chapter 1*, so we will not repeat ourselves. Instead, it will be interesting to show you a new useful example that we haven't yet covered; this example sets two (or more) path parameters:

```
app.get('/:userId/pets/:petId', function getPet (request,
reply) {
  reply.send(request.params)
})
```

The `request.params` object contains both parameters, `userId` and `petId`, which are declared in the URL string definition.

The last thing to know about the path parameters is the maximum length they might have. By default, an URL parameter can't be more than 100 characters. This is a security check that Fastify sets by default, and that you can customize in the root server instance initialization:

```
const app = fastify({
  maxParamLength: 40
})
```

Since all your application's path parameters should not exceed a known length limit, it is good to reduce it. Consider that it is a global setting, and you can't change it for a single route.

If a client hits the parameter length limit, it will get a 404 Not Found response.

The query string

The query string is an additional part of the URL string that the client can append after a question mark:

```
http://localhost:8080/foo/bar?param1=one&param2=two
```

These params let your clients submit information to those endpoints that don't support the request payload, such as `GET` or `HEAD HTTP`. Note that it is possible to retrieve only input strings and not complex data such as a file.

To read this information, you can access the `request.query` field wherever you have the `Request` component: hooks, decorators, or handlers.

Fastify supports basic 1:1 relation mapping, so a `foo.bar=42` query parameter produces a `{ "foo.bar" : "42" }` query object. Meanwhile, we should expect a nested object like this:

```
{
  "foo": {
    "bar": "42"
```

```
  }
}
```

To do so, we must change the default query string parser with qs, a new external module, (https://www.npmjs.com/package/qs):

```
const qs = require('qs')
const app = fastify({
  querystringParser: function newParser (queryParamString)
  {
    return qs.parse(queryParamString, { allowDots: true })
  }
})
```

This setup unlocks a comprehensive set of new syntaxes that you can use in query strings such as arrays, nested objects, and custom char separators.

The headers

The headers are a key-value map that can be read as a JavaScript object within the request.headers property. Note that, by default, Node.js will apply a lowercase format to every header's key. So, if your client sends to your Fastify server the CustomHeader: AbC header, you must access it with the request.headers.customheader statement.

This logic follows the HTTP standard that stands for case-insensitive field names.

If you need to get the original headers sent by the client, you must access the request.raw.rawHeaders property. Consider that request.raw gives you access to the Node.js http.IncomingMessage object, so you are free to read data added to the Node.js core implementation, such as the raw headers.

The body

The request's body can be read through the request.body property. Fastify handles two input content types:

1. The application/json produces a JavaScript object as a body value
2. The text/plain produces a string that will be set as a request.body value

Note that the request payload will be read for the POST, PUT, PATCH, OPTIONS, and DELETE HTTP methods. The GET and HEAD ones don't parse the body, as per the HTTP/1.1 specification.

Fastify sets a length limit to the payload to protect the application from **Denial-of-Service** (**DOS**) attacks, sending a huge payload to block your server in the parsing phase. When a client hits the default 1-megabyte limit, it receives a **413 - Request body is too large error** response. For example, this could be an unwanted behavior during an image upload. So, you should customize the default body size limit by setting the options as follows:

```
const app = fastify({ // [1]
  bodyLimit: 1024 // 1 KB
})
app.post('/', { // [2]
  bodyLimit: 2097152 // 2 MB
}, handler)
```

The **[1]** configuration defines the maximum length for every route without a custom limit, such as route **[2]**.

> **Security first**
>
> It is a good practice to limit the default body size to the minimum value you expect, and to set a specific limit for routes that need more input data. Usually, 256 KB is enough for simple user input.

The user input is not JSON and text only. We will discuss how to avoid body parsing or manage more content types such as `multipart/form-data` in *Chapter 4*.

We have covered the route configuration and learned how to read the HTTP request input sources. Now we are ready to take a deeper look at the routes' organization into plugins!

Managing the route's scope

In Fastify, an endpoint has two central aspects that you will set when defining a new route:

1. The route configuration
2. The server instance, where the route has been registered

This metadata controls how the route behaves when the client calls it. Earlier in this chapter, we saw the first point, but now we must deepen the second aspect: the server instance context. The **route's scope** is built on top of the server's instance context where the entry point has been registered. Every route has its own route scope that is built during the startup phase, and it is like a settings container that tracks the handler's configuration. Let's see how it works.

The route server instance

When we talk about the **route's scope**, we must consider the server instance where the route has been added. This information is important because it will define the following:

- The handler execution context

- The request life cycle events

- The default route configuration

The product of these three aspects is the route's scope. The route's scope cannot be modified after application startup, since it is an optimized object of all the components that must serve the HTTP requests.

To see what this means in practice, we can play with the following code:

```
app.get('/private', function handle (request, reply) {
  reply.send({ secret: 'data' })
})
app.register(function privatePlugin (instance, opts, next) {
  instance.addHook('onRequest', function isAdmin (request,
  reply, done) {
    if (request.headers['x-api-key'] === 'ADM1N') {
      done()
    } else {
      done(new Error('you are not an admin'))
    }
  })
  next()
})
```

By calling the `http://localhost:8080/private` endpoint, the `isAdmin` hook **will never be executed because the route is defined in the app scope**. The `isAdmin` hook is declared in the `privatePlugin`'s context only.

Moving the `/private` endpoint declaration into the `privatePlugin` context, and taking care of changing the `app.get` code to `instance.get`, will change the route's server instance context. Restarting the application and making a new HTTP request will execute the `isAdmin` function because the route's scope has changed.

We have explored this aspect of the framework in *Chapter 2*, where we learned how the encapsulation scope affects the Fastify instances registered. In detail, I'm referring to **decorators and hooks**: a route that inherits all the **request life cycle hooks** registered in the server instance it belongs to, and the decorators too.

Consequently, the server's instance context impacts all the routes added in that scope and its children, as shown in the previous example.

To consolidate this aspect, we can take a look at another example with more plugins:

```
app.addHook('onRequest', function parseUserHook (request,
reply, done) {
  const level = parseInt(request.headers.level) || 0
  request.user = {
    level,
    isAdmin: level === 42
  }
  done()
})
app.get('/public', handler)
app.register(rootChildPlugin)
async function rootChildPlugin (plugin) {
  plugin.addHook('onRequest', function level99Hook (request
  ,reply, done) {
    if (request.user.level < 99) {
      done(new Error('You need an higher level'))
      return
    }
    done()
  })
  plugin.get('/onlyLevel99', handler)
  plugin.register(childPlugin)
}
async function childPlugin (plugin) {
  plugin.addHook('onRequest', function adminHook (request,
  reply, done) {
    if (request.user.isAdmin) {
      done(new Error('You are not an admin'))
      return
    }
    done()
  })

  plugin.get('/onlyAdmin', handler)
}
```

Take some time to read the previous code carefully; we have added one route and one `onRequest` hook into each context:

- The `/public` route in the root `app` application instance
- The `/onlyLevel99` URL in the `rootChildPlugin` function, which is the `app` context's child
- The `/onlyAdmin` endpoint in the `childPlugin` context is registered in the `rootChildPlugin` function

Now, if we try to call the `/onlyAdmin` endpoint, the following will happen:

1. The server receives the HTTP request and does the routing process, finding the right handler.
2. The handler is registered in the `childPlugin` context, which is a child server instance.
3. Fastify traverses the context's tree till the root application instance and starts the **request life cycle** execution.
4. Every hook in the traversed contexts is executed sequentially. So, the executive order will be as follows:

   a. The `parseUserHook` hook function adds a user object to the HTTP request.

   b. The `level99Hook` will check whether the user object has the appropriate level to access the routes defined in that context and its children's context.

   c. The `adminHook` finally checks whether the user object is an admin.

It is worth repeating that if the `/onlyAdmin` route was registered in the `app` context, the fourth point would only execute the hooks added to that context.

In our examples, we used hooks, but the concept would be the same for decorators: a decorator added in the `rootChildPlugin` context is not available to be used in the app's context because it is a parent. Instead, the decorator will be ready to use in the `childPlugin` context because it is a child of `rootChildPlugin`.

The route's context is valuable because a route can access a database connection or trigger an authentication hook only if it has been added to the right server instance. For this reason, knowing in which context a route is registered is very important. There is a set of debugging techniques you can use to understand where a route is registered, which we will examine later.

Printing the routes tree

During the development process, especially when developing the first Fastify application, you might feel overwhelmed by having to understand the functions executed when a request reaches an endpoint. Don't panic! This is expected at the beginning, and the feeling is temporary.

To reduce the stress, Fastify has a couple of debugging outputs and techniques that are helpful to unravel a complex code base.

The utilities we are talking about can be used like so:

```
app.ready()
  .then(function started () {
    console.log(app.printPlugins()) // [1]
    console.log(app.printRoutes({ commonPrefix: false }))
    // [2]
  })
```

The printPlugins() method of **[1]** returns a string with a tree representation containing all the encapsulated contexts and loading times. The output gives you a complete overview of all the plugins created and the nesting level. Instead, the printRoutes() method of **[2]** shows us the application's routes list that we are going to see later.

To view an example of the printPlugins function, consider the following code:

```
app.register(async function pluginA (instance) {
  instance.register(async function pluginB (instance) {
    instance.register(function pluginC (instance, opts,
    next) {
      setTimeout(next, 1000)
    })
  })
})
app.register(async function pluginX () {})
```

We have created four plugins: three nested into each other and one at the root context. The printPlugins executions produce the following output string:

```
bound root 1026 ms
├── bound _after 3 ms
├┬ pluginA 1017 ms
| └┬ pluginB 1017 ms
|   └── pluginC 1016 ms
└── pluginX 0 ms
```

Here, we can see two interesting things:

1. The names in the output are the plugins' function names: This reaffirms the importance of preferring named functions instead of anonymous ones. Otherwise, the debugging phase will be more complicated.

2. The loading times are cumulative: The root time loading is the sum of all its children's contexts. For this reason, the pluginC loading time impacts the parent ones.

This output helps us to do the following:

- Get a complete overview of the application tree. In fact, adding a new route to the `pluginB` context inherits the configuration of that scope and the parent ones.

- Identify the slower plugins to load.

> **Fastify internals**
>
> The output shows the `bound _after` string. You can simply ignore this string output because it is an internal Fastify behavior, and it does not give us information about our application.

Looking at the `printRoutes()` method of **[2]** at the beginning of this section, we can get a complete list of all the routes that have been loaded by Fastify. This helps you to get an easy-to-read output tree:

```
└── / (GET)
    ├── dogs (GET, POST, PUT)
    │   └── /:id (GET)
    ├── feline (GET)
    │   ├── / (GET)
    │   └── /cats (GET, POST, PUT)
    │       /cats (GET) {"host":"cats.land"}
    │       └── /:id (GET)
    └── who-is-the-best (GET)
```

As you can see, the print lists all the routes within their HTTP methods and constraints.

As you may remember, in the `printRoutes()` **[2]** statement, we used the `commonPrefix` option. This is necessary to overcome the internal Radix-tree we saw in the previous *Routing to the endpoint* section. Without this parameter, Fastify will show you the internal representation of the routes. This means that the routes are grouped by the most common URL string. The following set of routes has three routes with the `hel` prefix in common:

```
app.get('/hello', handler)
app.get('/help', handler)
app.get('/:help', handler)
app.get('/helicopter', handler)
app.get('/foo', handler)
app.ready(() => { console.log(app.printRoutes()) })
```

Printing those routes by calling the `printRoutes()` function produces the following:

```
└── /
    ├── hel
    │   ├── lo (GET)
    │   ├── p (GET)
```

```
|      └── icopter (GET)
├── :help (GET)
└── foo (GET)
```

As the preceding output confirms, the `hel` string is the most shared URL string that groups three routes. Note that the `:help` route is not grouped: this happens because it is a path parameter and not a static string. As mentioned already, this output shows the router's internal logic, and it may be hard to read and understand. Going deeper into the route's internal details is beyond the scope of this book because it concerns the internal Radix-tree we mentioned in the *Routing to the endpoint* section.

The `printRoutes()` method supports another useful option flag: `includeHooks`. Let's add the following Boolean:

```
app.printRoutes({ commonPrefix: false, includeHooks: true })
```

This will print to the output tree all the hooks that the route will execute during the **request life cycle**! The print is extremely useful to spot hooks that run when you would not expect them!

To set an example, let's see the following sample code:

```
app.addHook('preHandler', async function isAnimal () { })
app.get('/dogs', handler)
app.register(async function pluginA (instance) {
  instance.addHook('onRequest', async function isCute () { })
  instance.addHook('preHandler', async function isFeline () { })
  instance.get('/cats', {
    onRequest: async function hasClaw () { } // [1]
  }, handler)
})
await app.ready()
console.log(app.printRoutes({ commonPrefix: false,
includeHooks: true }))
```

Print out the string:

```
└── / (-)
    ├── cats (GET)
    |    • (onRequest) ["isCute()","hasClaw()"]
    |    • (preHandler) ["isAnimal()","isFeline()"]
    └── dogs (GET)
         • (preHandler) ["isAnimal()"]
```

The output tree is immediately readable, and it is telling us which function runs for each hook registered! Moreover, the functions are ordered by execution, so we are sure that `isFeline` runs after the `isAnimal` function. In this snippet, we used the route hook registration [1] to highlight how the hooks append each other in sequence.

> **Named functions**
>
> As mentioned in the *Adding basic routes* section of *Chapter 1*, using arrow functions to define the application's hooks will return the `"anonymous ()"` output string. This could hinder you from debugging the route

Now you can literally see your application's routes! You can use these simple functions to get an overview of the plugin's structure and gain a more detailed output of each endpoint, to understand the flow that an HTTP request will follow.

> **Drawing a schema image of your application**
>
> By using the `fastify-overview` plugin from `https://github.com/Eomm/fastify-overview`, you will be able to create a graphical layout of your application with all the hooks, decorators, and Fastify-encapsulated contexts highlighted! You should give it a try.

Get ready for the next section, which will introduce more advanced topics, including how to start working with hooks and route configurations together.

Adding new behaviors to routes

At the beginning of this chapter, we learned how to use the `routeOptions` object to configure a route, but we did not talk about the `config` option!

This simple field gives us the power to do the following:

- Access the config in the handler and hook functions
- Implement the **Aspect-Oriented Programming (AOP)** that we are going to see later

How does it work in practice? Let's find out!

Accessing the route's configuration

With the `routerOption.config` parameter, you can specify a JSON that contains whatever you need. Then, it is possible to access it later within the `Request` component in the handlers or hooks' function through the `context.config` field:

```
async function operation (request, reply) {
  return request.context.config
}
app.get('/', {
  handler: operation,
  config: {
    hello: 'world'
```

```
    }
  })
```

In this way, you can create a business logic that depends on modifying the components' behavior.

For example, we can have a `preHandler` hook that runs before the `schedule` handler function in each route:

```
app.addHook('preHandler', async function calculatePriority
(request) {
  request.priority = request.context.config.priority    10
})
app.get('/private', {
  handler: schedule,
  config: { priority: 5 }
})
app.get('/public', {
  handler: schedule,
  config: { priority: 1 }
})
```

The `calculatePriority` hook adds a level of priority to the request object, based on the route configuration: the `/public` URL has less importance than the `/private` one.

By doing so, you could have generic components: handlers or hooks that act differently based on the route's options.

AOP

The AOP paradigm focuses on isolating cross-cutting concerns from the business logic and improving the system's modularity.

To be less theoretical and more practical, AOP in Fastify means that you can isolate boring stuff into hooks and add it to the routes that need it!

Here is a complete example:

```
app.addHook('onRoute', function hook (routeOptions) {
  if (routeOptions.config.private === true) {
    routeOptions.onRequest = async function auth (request)
    {
      if (request.headers.token !== 'admin') {
        const authError = new Error('Private zone')
        authError.statusCode = 401
        throw authError
      }
    }
```

```
  }
})
app.get('/private', {
  config: { private: true },
  handler
})
app.get('/public', {
  config: { private: false },
  handler
})
```

Since the first chapter, we have marginally discussed hooks. In the code, we introduced another hook example: the onRoute application hook, which listens for every route registered in the app context (and its children contexts). It can mutate the routeOptions object before Fastify instantiates the route endpoint.

The routes have config.private fields that tell the hook function to add an onRequest hook to the endpoint, which is only created if the value is true.

Let's list all the advantages here:

- You can isolate a behavior into a plugin that adds hooks to the routes only if needed.

- Adding a generic hook function that runs when it is not necessary consumes resources and reduces the overall performance.

- A centralized registration of the application's crucial aspects, which reduces the route's configuration verbosity. In a real-world application, you will have more and more APIs that you will need to configure.

- You can keep building plugins with company-wide behavior and reuse them across your organization.

This example shows you the power of the onRoute hook in conjunction with the routeOptions. config object. In future chapters, you will see other use cases that are going to leverage this creational middleware pattern to give you more tools and ideas to build applications faster than ever.

We have just seen how powerful the route's config property is and how to use it across the application.

Summary

This chapter has explained how routing works in Fastify, from route definition to request evaluation.

Now, you know how to add new endpoints and implement an effective handler function, both async or sync, with all the different aspects that might impact the request flow. You know how to access the client's input to accomplish your business logic and reply effectively with a success or an error.

We saw how the server context could impact the route handler implementation, executing the hooks in that encapsulated context and accessing the decorators registered. Moreover, you learned how to tweak the route initialization by using the `onRoute` hook and the route's `config`: using Fastify's features together gives us new ways to build software even more quickly!

The routing has no more secrets for you, and you can define a complete set of flexible routes to evolve thanks to the constraints and manage a broad set of real-world use cases to get things done.

In the next chapter, we will discuss, in detail, one of Fastify's core concepts, which we have already mentioned briefly and seen many times in our examples: hooks and decorators!

4

Exploring Hooks

In this chapter, we will learn what makes Fastify different from the majority of other web frameworks. In fact, contrary to several other frameworks that have the concept of middleware, Fastify is based on **hooks**. They all have different use cases, and mastering their usage is key to developing stable and production-ready server applications.

Even if the topic is considered somehow complex, the goal of the subsequent sections is to give a good overview of how the Fastify framework "thinks" and build the right confidence in using these tools.

Before going into their details, though, we need to introduce another concept that makes everything even possible. Namely, we need to learn about the **lifecycle** of Fastify applications.

In this chapter, we will focus on these concepts:

- What is a lifecycle?
- Declaring hooks
- Understanding the application lifecycle
- Understanding the request and reply lifecycle

Technical requirements

To follow this chapter, you will need the following:

- A text editor, such as VS Code
- A working Node.js v18 installation
- Access to a shell such as Bash or CMD
- The `curl` command-line application

All the code snippets for this chapter are available on GitHub at `https://github.com/PacktPublishing/Accelerating-Server-Side-Development-with-Fastify/tree/main/Chapter%204`.

What is a lifecycle?

When talking about a lifecycle in the context of a framework, we refer to the order in which the functions are executed. The tricky part is that application developers write only a subset of its code while the rest is developed and bundled by the framework developers.

During the application execution, the function calls bounce between internal and user-written code, and it might be hard to follow what is going on. So, it comes naturally at this point that having deep knowledge of this subject is crucial.

The lifecycle depends on the architecture of the framework. Two well-known architectures are usually used to develop a web framework:

- **Middleware-based architecture**: Thanks to its more manageable learning curve, this is the most known lifecycle architecture, but time has proved it to have some significant drawbacks. When working with this architecture, an application developer must care about the order of declaration of the middleware functions since it influences the order of execution at runtime. Of course, it can be hard to track the execution order in bigger and more complex applications across multiple modules and files. Moreover, writing reusable code might be more challenging because every moving part is more tidily coupled. As a final note, every middleware function will be executed at every cycle, whether needed or not.

- **Hook-based architecture**: Fastify, unlike some other famous frameworks, implements a hook-based one. This kind of architecture was a precise design decision from day one since a hook system is more scalable and easier to maintain in the long run. As a nice bonus, since a hook only executes if needed, it usually leads to faster applications! However, it is worth mentioning that it is also harder to master. As we already briefly mentioned, we have different and specific components to deal with in a hook-based architecture.

For the rest of this chapter, we will talk in detail about hook-based architecture.

When dealing with web frameworks, there are usually at least two main lifecycles:

- **The application lifecycle**: This lifecycle is in charge of the different phases of the application server execution. It mainly deals with the server boot sequence and shutdown. We can attach "global" functionalities that are shared between every route and plugin. Moreover, we can act at a specific moment of the lifecycle execution, adding a proper hook function. The most common actions we perform are after the server is started or before it shuts down.

- **The request/reply lifecycle**: The request/response phase is the core of every client-server application. Almost the entirety of the execution time is spent inside this sole sequence. For this reason, this lifecycle usually has way more phases and therefore hooks we can add to it. The most common ones are content parsers, serializers, and authorization hooks.

Now that we understand more about the lifecycle types, we can spend the rest of the chapter learning how Fastify implements them.

Declaring hooks

In the previous section, we saw that a server-side application usually has two main lifecycles. So, being a hook-based web framework and following its philosophy of giving complete control to developers, Fastify emits a specific event every time it advances to the next phase. Furthermore, these phases follow a rigid and well-defined execution order. Knowing it enables us to add functionality at a specific point during the boot or the execution of our application.

One essential and beneficial side effect of this approach is that, as developers, we don't care about the declaration order of our hooks since the framework guarantees that they will be invoked at the right moment in time.

The mechanism described works because Fastify, under the hood, defines a "hook runner" that runs the callback functions declared for every known event. As developers, we need a method to attach our hooks to the Fastify instance. The `addHook` hook allows us to do precisely that, besides being an application or request/reply hook.

> **Callback-based versus async hooks**
>
> In *Chapter 2*, we saw that we could declare plugins with two different styles: callback-based and async functions. The same applies here. Again, it is essential to choose one style and stick with it. Mixing them can lead to unexpected behavior. As already decided in this book, we will use only async functions. One last thing to remember is that some hooks are only synchronous. We will clarify this when speaking about them.

As we can see in the following `add-hook.cjs snippet` method, it takes two arguments:

- A name of the event we want to listen

- The callback function:

```
...
fastify.addHook('onRequest', (…) => {})
...
```

Here, we omitted the callback's signature and the return value since every hook has its own.

We can call addHook on the same event multiple times to declare more than one hook. In the following sections, we will learn all the events emitted by Fastify and describe every callback function in depth.

Understanding the application lifecycle

The application lifecycle covers the boot process and the execution of our application server. In particular, we refer to loading the plugins, adding routes, making the HTTP server run, and eventually closing it. Fastify will emit four different events, allowing us to interact with the behavior of every phase:

- onRoute
- onRegister
- onReady
- onClose

Now, for every event of the previous list, let's check the respective callback signature and most common use cases.

The onRoute hook

The onRoute hook event is triggered every time a route is added to the Fastify instance. This callback is a **synchronous function** that takes one argument, commonly called routeOptions. This argument is a mutable object reference to the route declaration object itself, and we can use it to modify route properties.

This hook is encapsulated and doesn't return any value. Therefore, one of the most common use cases is adding route-level hooks or checking for the presence of a specific option value and acting accordingly.

We can see a trivial example in on-route.cjs, where we add a custom route-level preHandler function to the route properties:

```
const Fastify = require('fastify')
const app = Fastify({ logger: true })

app.addHook('onRoute', routeOptions => { // [1]
  async function customPreHandler(request, reply) {
    request.log.info('Hi from customPreHandler!')
  }
  app.log.info('Adding custom preHandler to the route.')
  routeOptions.preHandler = [...(routeOptions.preHandler ??
  []), customPreHandler] // [2]
})
app.route({
  // [3]
```

```
    url: '/foo',
    method: 'GET',
    schema: {
      200: {
        type: 'object',
        properties: {
          foo: {
            type: 'string'
          }
        }
      }
    },
    handler: (req, reply) => {
      reply.send({ foo: 'bar' })
    },
  })

app
  .listen({ port: 3000 })
  .then(() => {
    app.log.info('Application is listening.')
  })
  .catch(err => {
    app.log.error(err)
    process.exit()
  })
```

After setting up the Fastify instance, we add a new onRoute hook ([**1**]). The sole purpose of this hook is to add a route-level preHandler hook ([**2**]) to the route definition, even if there weren't any previous hooks defined for this route ([**3**]).

Here, we can learn two important outcomes that will help us when dealing with route-level hooks:

- The routeOptions object is mutable, and we can change its properties. However, if we want to keep the previous values, we need to explicitly re-add them ([**2**]).

- Route-level hooks are arrays of hook functions (we will see more about this later in the chapter).

Now, we can check the output of the snippet by opening a new terminal and running the following command:

```
$ node on-route.cjs
```

This command will start our server on port 3000.

Now we can use `curl` in a different terminal window to request the server and check the result in the server console:

```
$ curl localhost:3000/foo
```

We can search for the message `"Hi from customPreHandler!"` in the logs to check whether our `customHandler` was executed:

```
{"level":30,"time":1635765353312,"pid":20344,"hostname":"localhost","r
eqId":"req-1","msg":"Hi from customPreHandler!"}
```

This example only scratches the surfaces of the possible use cases. We can find the complete definition of the `routeOptions` properties in the official documentation (`https://www.fastify.io/docs/latest/Reference/Routes/#routes-options`).

The onRegister hook

The `onRegister` hook is similar to the previous one regarding how it works, but we can use it when dealing with plugins. In fact, for every registered plugin that creates a new encapsulation context, the `onRegister` hooks are executed before the registration.

We can use this hook to discover when a new context is created and add or remove functionality; as we already learned, during the plugin's registration and thanks to its robust encapsulation, Fastify creates a new instance with a child context. Note that this hook's callback function won't be called if the registered plugin is wrapped in `fastify-plugin`.

The `onRegister` hook accepts a synchronous callback with two arguments. The first parameter is the newly created Fastify instance with its encapsulated context. The latter is the `options` object passed to the plugin during the registration.

The following `on-register.cjs` snippet shows an easy yet non-trivial example that covers encapsulated, and non-encapsulated plugins use cases:

```
const Fastify = require('fastify')
const fp = require('fastify-plugin')

const app = Fastify({ logger: true })
app.decorate('data', { foo: 'bar' }) // [1]
app.addHook('onRegister', (instance, options) => {
  app.log.info({ options })
  instance.data = { ...instance.data } // [2]
})
app.register(
  async function plugin1(instance, options) {
    instance.data.plugin1 = 'hi' // [3]
    instance.log.info({ data: instance.data })
```

```
    },
    { name: 'plugin1' }
  )
  app.register(
    fp(async function plugin2(instance, options) {
      instance.data.plugin2 = 'hi2' // [4]
      instance.log.info({ data: instance.data })
    }),
    { name: 'plugin2' }
  )

  app
    .ready()
    .then(() => {
      app.log.info('Application is ready.')
      app.log.info({ data: app.data }) // [5]
    })
    .catch(err => {
      app.log.error(err)
      process.exit()
    })
```

First, we decorate the top-level Fastify instance with a custom data object ([1]). Then we attach an onRegister hook that logs the options plugin and shallow-copy the data object ([2]). This will effectively create a new object that inherits the foo property, allowing us to have encapsulated the data object. At [3], we register our first plugin that adds the plugin1 property to the object. On the other hand, the second plugin is registered using fastify-plugin ([4]), and therefore Fastify will not trigger our onRegister hook. Here, we modify the data object again, adding the plugin2 property to it.

Shallow-copy versus deep copy

Since an object is just a reference to the allocated memory address in JavaScript, we can copy the objects in two different ways. By default, we "shallow-copy" them: if one source object's property references another object, the copied property will point to the same memory address. We implicitly create a link between the old and new property. If we change it in one place, it is reflected in the other. On the other hand, deep-copying an object means that whenever a property references another object, we will create a new reference and, therefore, a memory allocation. Since deep copying is expensive, all methods and operators included in JavaScript make shallow copies.

Let's execute this script in a terminal window and check the logs:

```
$ node on-register.cjs
{"level":30,"time":1636192862276,"pid":13381,"hostname":"localhost",
"options":{"name":"plugin1"}}
{"level":30,"time":1636192862276,"pid":13381,"hostname":"localhost",
"data":{"foo":"bar","plugin1":"hi"}}
{"level":30,"time":1636192862277,"pid":13381,"hostname":"localhost",
"data":{"foo":"bar","plugin2":"hi2"}}
{"level":30,"time":1636192862284,"pid":13381,"hostname":"localhost",
"msg":"Application is ready."}
{"level":30,"time":1636192862284,"pid":13381,"hostname":"localhost",
"data":{"foo":"bar","plugin2":"hi2"}}
```

We can see that adding a property in plugin1 hasn't any repercussions on the top-level data property. On the other hand, since plugin2 is loaded using fastify-plugin, it has the same context as the main Fastify instance ([5]), and the onRegister hook isn't even called.

The onReady hook

The onReady hook is triggered after fastify.ready() is invoked and before the server starts listening. If the call to ready is omitted, then listen will automatically call it, and these hooks will be executed anyway. Since we can define multiple onReady hooks, the server will be ready to accept incoming requests only after all of them are completed.

Contrary to the other two hooks we already saw, this one is asynchronous. Therefore, it is crucial to define it as an async function or manually call the done callback to progress with the server boot and code execution. In addition to this, the onReady hooks are invoked with the this value bound to the Fastify instance.

> **A bound this context**
>
> When dealing with Fastify functionalities that have an explicitly bound this value, as in the case of the onReady hook, it is essential to use the old function syntax instead of the arrow function one. Using the latter will prevent binding, making it impossible to access the instance and custom data added to it.

In the on-ready.cjs snippet, we show a straightforward example of the Fastify bound context:

```
const Fastify = require('fastify')

const app = Fastify({ logger: true })
app.decorate('data', 'mydata') // [1]
app.addHook('onReady', async function () {
  // [2]
  app.log.info(this.data)
```

```
})

app
  .ready()
  .then(() => {
    app.log.info('Application is ready.')
  })
  .catch(err => {
    app.log.error(err)
    process.exit()
  })
```

At [1], we decorate the primary instance with a dummy value. Then we add one onReady hook using the async function syntax. After that, we log the data value to show the bound this inside it ([2]).

Running this snippet will produce a short output:

```
{"level":30,"time":1636284966854,"pid":3660,"hostname":"localhost",
"msg":"mydata"}
{"level":30,"time":1636284966855,"pid":3660,"hostname":" localhost ",
"msg":"Application is ready."}
```

We can check that mydata is logged before the application is ready and that, indeed, we have access to the Fastify instance via this.

The onClose hook

While the hooks we learned about in the last three sections are used during the boot process, on the other hand, onClose is triggered during the shutdown phase right after fastify.close() is called. Thus, it is handy when plugins need to do something right before stopping the server, such as cleaning database connections. This hook is asynchronous and accepts one argument, the Fastify instance. As usual, when dealing with async functionalities, there is also an optional done callback (the second argument) if the async function isn't used.

In the on-close.cjs example, we choose to use the async function to log a message:

```
const Fastify = require('fastify')

const app = Fastify({ logger: true })
app.addHook('onClose', async (instance) => {
  // [1]
  instance.log.info('onClose hook triggered!')
})

app
  .ready()
```

```
.then(async () => {
  app.log.info('Application is ready.')
  await app.close() // [2]
})
.catch(err => {
  app.log.error(err)
  process.exit()
})
```

After adding a dummy `onClose` ([1]) hook, we explicitly call `app.close()` on [2] to trigger it.

After running the example, we can see that the last thing logged is indeed the hook line:

```
{"level":30,"time":1636298033958,"pid":4257,"hostname":"localhost",
"msg":"Application is ready."}
{"level":30,"time":1636298033959,"pid":4257,"hostname":"localhost ",
"msg":"onClose hook triggered!"}
```

With the `onClose` hook, we have finished our discussion about the application-level lifecycle. Now, we will move to the more numerous and, therefore, exciting request-reply hooks.

Understanding the request and reply lifecycle

When executing a Fastify server application, the vast majority of the time is spent in the request-reply cycle. As developers, we define routes that the clients will call and produce a response based on the incoming conditions. In true Fastify philosophy, we have several events at our disposal to interact with this cycle. As usual, they will be triggered automatically by the framework only when needed. These hooks are fully encapsulated so that we can control their execution context with the `register` method.

As we saw in the previous section, we had four application hooks. Here, we have nine request and reply hooks:

- `onRequest`
- `preParsing`
- `preValidation`
- `preHandler`
- `preSerialization`
- `onSend`
- `onResponse`
- `onError`
- `onTimeout`

Since they are part of the request/reply cycle, the trigger order of these events is crucial. Therefore, the first seven elements of the list are written from the first to the last. Furthermore, onError and onTimeout can be triggered in no specific order at every step in the cycle since an error or a timeout can happen at any point.

Let's take a look at the following image to understand the Fastify request/reply lifecycle better:

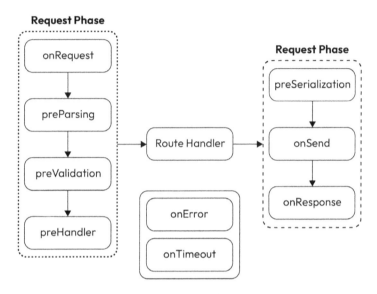

Figure 4.1: The request and reply lifecycle

For clarity, we divided the hooks into groups using bubbles. As we already said, there are three main groups.

Inside the dotted bubble, we can see the request phase. These callback hooks are called, in top-down order, following the arrow's direction, before the user-defined handler for the current route.

The dashed bubble contains the reply phase, whose hooks are called after the user-defined route handler. Every hook, at every point, can throw an error or go into the timeout. If it happens, the request will finish in the solid bubble, leaving the normal flow.

We will learn more about request and reply hooks in the subsequent sections, starting from error handling.

Handling errors inside hooks

During the hook execution, an error can occur. Since hooks are just asynchronous callback functions automatically invoked by the framework, error handling follows the same rules for standard JavaScript functions. Again, the only main difference is the two styles defining them.

If we choose the callback style (remember, in this book, we are only using the async function style), we have to pass the error manually to the done callback:

```
fastify.addHook('onRequest', (request, reply, done) => {
  done(new Error('onRequest error'))
})
```

On the other hand, when declaring an async function, it is enough to throw the error:

```
fastify.addHook('onResponse', async (request, reply) => {
  throw new Error('Some error')
})
```

Since we have access to the reply object, we can also change the reply's response code and reply to the client directly from the hook. If we choose not to reply in the case of an error or to reply with an error, then Fastify will call the onError hook.

Now that we understand how errors modify the request-reply flow, we can finally start analyzing every hook Fastify puts at our disposal.

The onRequest hook

As the name implies, the onRequest hook is triggered every time there is an incoming request. Since it is the first event on the execution list, the body request is always null because body parsing doesn't happen yet. The hook function is asynchronous and accepts two parameters, the Request and Reply Fastify objects.

Besides showing how onRequest works, the following on-request.cjs snippet also shows how the hook encapsulation works (as we already said, the encapsulation is valid for every other hook as well):

```
const Fastify = require('fastify')

const app = Fastify({ logger: true })
app.addHook('onRequest', async (request, reply) => {
  // [1]
  request.log.info('Hi from the top-level onRequest hook.')
})
app.register(async function plugin1(instance, options) {
  instance.addHook('onRequest', async (request, reply) => {
    // [2]
    request.log.info('Hi from the child-level onRequest
      hook.')
  })

  instance.get('/child-level', async (_request, _reply) =>
```

```
  {
    // [3]
    return 'child-level'
  })
})
app.get('/top-level', async (_request, _reply) => {
  // [4]
  return 'top-level'
})

app.listen({ port: 3000 }).catch(err => {
  app.log.error(err)
  process.exit()
})
```

The first thing we do is add a top-level onRequest hook ([1]). Then we register a plugin that defines another onRequest hook ([2]) and a GET /child-level route ([3]). Finally, we add another GET route on the /top-level path ([4]).

Let's run the script in the terminal and check the output:

```
$ node on-request.cjs
{"level":30,"time":1636444514061,"pid":30137,"hostname":
"localhost","msg":"Server listening at http://127.0.0.1:3000"}
```

This time our server is running on port 3000, and it is waiting for incoming connections.

We can open another terminal and use curl to make our calls.

First of all, let's get the /child-level route:

```
$ curl localhost:3000/child-level
child-level
```

We can see that curl was able to get the route response correctly. Switching the terminal window again to the one that is running the server, we can check the logs:

```
{"level":30,"time":1636444712019,"pid":30137,"hostname":"localhost",
"reqId":"req-1","req":{"method":"GET","url":"/child-level","hostname":
"localhost:3000","remoteAddress":"127.0.0.1","remotePort":56903},
"msg":"incoming request"}
{"level":30,"time":1636444712021,"pid":30137,"hostname":" localhost ",
"reqId":"req-1","msg":"Hi from the top-level onRequest hook."}
{"level":30,"time":1636444712023,"pid":30137,"hostname":" localhost ",
"reqId":"req-1","msg":"Hi from the child-level onRequest hook."}
{"level":30,"time":1636444712037,"pid":30137,"hostname":" localhost ",
"reqId":"req-1","res":{"statusCode":200},"responseTime"
:16.760624945163727,"msg":"request completed"}
```

We can spot immediately that both hooks were triggered during the request-response cycle. Moreover, we can also see the order of the invocations: first, `"Hi from the top-level onRequest hook"`, and then `"Hi from the child-level onRequest hook"`.

To ensure that hook encapsulation is working as expected, let's call the `/top-level` route. We can switch again to the terminal where `curl` ran and type the following command:

```
$ curl localhost:3000/top-level
top-level
```

Now coming back to the server log output terminal, we can see the following:

```
{"level":30,"time":1636444957207,"pid":30137,"hostname":"
localhost","reqId":"req-2","req":{"method":"GET","url":"/
top-level","hostname":"localhost:3000","remoteAddress":"127.0.0.1",
"remotePort":56910},"msg":"incoming request"}
{"level":30,"time":1636444957208,"pid":30137,"hostname":
" localhost ","reqId":"req-2","msg":"Hi from the top-level onRequest
hook."}
{"level":30,"time":1636444957210,"pid":30137,"hostname":" localhost
","reqId":"req-2","res":{"statusCode":200},"responseTime":
1.8535420298576355,"msg":"request completed"}
```

This time, the logs show that only the top-level hook was triggered. This is an expected behavior, and we can use it to add hooks scoped to specific plugins only.

The preParsing hook

Declaring a `preParsing` hook allows us to transform the incoming request payload before it is parsed. This callback is asynchronous and accepts three parameters: `Request`, `Reply`, and the payload stream. Again, the `body` request is null since this hook is triggered before `preValidation`. Therefore, we must return a stream if we want to modify the incoming payload. Moreover, developers are also in charge of adding and updating the `receivedEncodedLength` property of the returned value.

The `pre-parsing.cjs` example shows how to change the incoming payload working directly with streams:

```
const Fastify = require('fastify')
const { Readable } = require('stream')

const app = Fastify({ logger: true })
app.addHook('preParsing', async (request, _reply, payload)
=> {
  let body = ''
  for await (const chunk of payload) {
    // [1]
    body += chunk
```

```
  }
  request.log.info(JSON.parse(body)) // [2]

  const newPayload = new Readable() // [3]
  newPayload.receivedEncodedLength =
    parseInt(request.headers['content-length'], 10)
  newPayload.push(JSON.stringify({ changed: 'payload' }))
  newPayload.push(null)

  return newPayload
})
app.post('/', (request, _reply) => {
  request.log.info(request.body) //[4]
  return 'done'
})

app.listen({ port: 3000 }).catch(err => {
  app.log.error(err)
  process.exit()
})
```

At [1], we declare our preParsing hook that consumes the incoming payload stream and creates a body string. We then parse the body ([2]) and log the content to the console. At [3], we create a new Readable stream, assign the correct receivedEncodedLength value, push new content into it, and return it. Finally, we declare a dummy route ([4]) to log the body object.

Running the script in a terminal will start our server on port 3000:

```
$ node pre-parsing.cjs
{"level":30,"time":1636532643143,"pid":38684,"hostname":
"localhost","msg":"Server listening at http://127.0.0.1:3000"}
```

Now in another terminal window, we can use curl to call the route and check the logs:

```
$ curl --header "Content-Type: application/json" \
  --request POST \
  --data '{"test":"payload"}' localhost:3000
done
```

Returning to the server terminal, we can see this output:

```
{"level":30,"time":1636534552994,"pid":39232,"hostname":"localhost",
"reqId":"req-1","req":{"method":"POST","url":"/","hostname":
"localhost:3000","remoteAddress":"127.0.0.1","remotePort":58837},
"msg":"incoming request"}
{"level":30,"time":1636534553005,"pid":39232,"hostname":" localhost
","reqId":"req-1","test":"payload"}
{"level":30,"time":1636534553010,"pid":39232,"hostname":" localhost
```

```
","reqId":"req-1","changed":"payload"}
{"level":30,"time":1636534553018,"pid":39232,"hostname":" localhost
","reqId":"req-1","res":{"statusCode":200},"responseTime":23.695625007
152557,"msg":"request completed"}
```

Those logs show us how the payload changed after the `preParsing` call. Now, looking back at the `pre-parsing.cjs` snippet, the first call to the logger ([2]) logged the original body we sent from `curl`, while the second call ([3]) logged the `newPayload` content.

The preValidation hook

We can use the `preValidation` hook to change the incoming payload before it is validated. Since the parsing has already happened, we finally have access to the `body` request, which we can modify directly.

The hook receives two arguments, `request` and `reply`. In the `pre-validation.cjs` snippet, we can see that the callback is asynchronous and doesn't return any value:

```
const Fastify = require('fastify')

const app = Fastify({ logger: true })
app.addHook('preValidation', async (request, _reply) => {
  request.body = { ...request.body, preValidation: 'added' }
})
app.post('/', (request, _reply) => {
  request.log.info(request.body)
  return 'done'
})

app.listen({ port: 3000 }).catch(err => {
  app.log.error(err)
  process.exit()
})
```

The example adds a simple `preValidation` hook that modifies the parsed `body` object. Inside the hook, we use the spread operator to add a property to the body, and then we assign the new value to the `request.body` property again.

As usual, we can start our server in a terminal window:

```
$ node pre-validation.cjs
{"level":30,"time":1636538075248,"pid":39965,"hostname":"localhost",
"msg":"Server listening at http://127.0.0.1:3000"}
```

Then, after opening a second terminal,, we can make the same call we did for the preParsing hook:

```
$ curl --header "Content-Type: application/json" \
  --request POST \
  --data '{"test":"payload"}' localhost:3000
```

In the server output we can see that our payload is changed:

```
"level":30,"time":1636538082315,"pid":39965,"hostname":"localhost","r
eqId":"req-1","req":{"method":"POST","url":"/","hostname":"localhost:
3000","remoteAddress":"127.0.0.1","remotePort":59225},"msg":"incoming
request"}
{"level":30,"time":1636538082326,"pid":39965,"hostname":" localhost
","reqId":"req-1","test":"payload","preValidation":"added"}
{"level":30,"time":1636538082338,"pid":39965,"hostname":" localhost
","reqId":"req-1","res":{"statusCode":200},"responseTime":22.288374960
422516,"msg":"request completed"}
```

The preHandler hook

The preHandler hook is an async function that receives the request and the reply as its arguments, and it is the last callback invoked before the route handler. Therefore, at this point of the execution, the request.body object is fully parsed and validated. However, as we can see in pre-handler. cjs example, we can still modify the body or query values using this hook to perform additional checks or request manipulation before executing the handler:

```
const Fastify = require('fastify')

const app = Fastify({ logger: true })
app.addHook('preHandler', async (request, reply) => {
  request.body = { ...request.body, prehandler: 'added' }
  request.query = { ...request.query, prehandler: 'added' }
})
app.post('/', (request, _reply) => {
  request.log.info({ body: request.body })
  request.log.info({ query: request.query })
  return 'done'
})

app.listen({ port: 3000 }).catch(err => {
  app.log.error(err)
  process.exit()
})
```

Usually, this is the most used hook by developers, but it shouldn't be the case. More often than not, other hooks that come before `preHandler` are better suited for the vast majority of purposes. For example, frequently, the body doesn't need to be fully parsed and validated before performing our actions on the incoming request. Instead, we should use `preHandler` only when accessing or manipulating validated body properties.

Since this last snippet doesn't add anything new, we are omitting the output of running it. If needed, the same steps we used to run `pre-validation.cjs` can also be used here.

The preSerialization hook

The `preSerialization` hook is in the group of three hooks that are called after the route handler. The other two are `onSend` and `onResponse`, and we will cover them in the following sections.

Let's focus on the first one here. Since we are dealing with the response payload, `preSerialization` has a similar signature to the `preParsing` hook. It accepts a `request` object, a `reply` object, and a third payload parameter. We can use its return value to change or replace the response object before serializing and sending it to the clients.

There are two essential things to remember about this hook:

- It is not called if the payload argument is `string`, `Buffer`, `stream`, or `null`
- If we change the payload, it will be changed for every response, including errored ones

The following `pre-serialization.cjs` example shows how we can add this hook to the Fastify instance:

```
const Fastify = require('fastify')

const app = Fastify({ logger: true })
app.addHook('preSerialization', async (request, reply, payload) => {
  return { ...payload, preSerialization: 'added' }
})
app.get('/', (request, _reply) => {
  return { foo: 'bar' }
})

app.listen({ port: 3000 }).catch(err => {
  app.log.error(err)
  process.exit()
})
```

Inside the hook, we use the spread operator to copy the original payload and then add a new property. At this point, it is just a matter of returning this newly created object to modify the body before returning it to the client.

Let's run the snippet from the terminal:

```
$ node pre-serialization.cjs
{"level":30,"time":1636709732482,"pid":60009,"hostname":"localhost",
"msg":"Server listening at http://127.0.0.1:3000"}
```

Now, as usual, in a different terminal window, we can use `curl` to do our call to the server:

```
$ curl localhost:3000
{"foo":"bar","preSerialization":"added"}
```

The onSend hook

The `onSend` hook is the last hook invoked before replying to the client. Contrary to `preSerialization`, the `onSend` hook receives a payload that is already serialized. Moreover, it is always called, no matter the type of response payload. Even if it is harder to do, we can use this hook to change our response too, but this time we have to return one of `string`, `Buffer`, `stream`, or `null`. Finally, the signature is identical to `preSerialization`.

Let's make an example in the `on-send.cjs` snippet with the most straightforward payload, a string:

```
const Fastify = require('fastify')

const app = Fastify({ logger: true })
app.addHook('onSend', async (request, reply, payload) => {
  const newPayload = payload.replace('foo', 'onSend')
  return newPayload
})
app.get('/', (request, _reply) => {
  return { foo: 'bar' }
})

app.listen({ port: 3000 }).catch(err => {
  app.log.error(err)
  process.exit()
})
```

Since the payload is already serialized as a `string`, we can use the `replace` method to modify it before returning it to the client.

We can use the same `curl` method we already know to check whether the returned payload has `foo` replaced with `onSend`.

The onResponse hook

The onResponse hook is the last hook of the request-reply lifecycle. This callback is called after the reply has already been sent to the client. Therefore, we can't change the payload anymore. However, we can use it to perform additional logic, such as calling external services or collecting metrics. It takes two arguments, `request` and `reply`, and doesn't return any value. We can see a concrete example in `on-response.cjs`:

```
const Fastify = require('fastify')

const app = Fastify({ logger: true })
app.addHook('onResponse', async (request, reply) => {
  request.log.info('onResponse hook') // [1]
})
app.get('/', (request, _reply) => {
  return { foo: 'bar' }
})

app.listen({ port: 3000 }).catch(err => {
  app.log.error(err)
  process.exit()
})
```

The example is, again, straightforward. We add a dummy onResponse hook that prints a line to the log output. Running it using the usual method will show the log.

On the other hand, since inside the `on-response.cjs` snippet, we changed our code slightly and made the hook call `reply.send()`; we have a very different result:

```
app.addHook('onResponse', async (request, reply) => {
  reply.send('onResponse')
})
```

Even if the client receives the response correctly, our server will throw the `"Reply was already sent."` error. The error doesn't affect the request-response cycle, and it is only printed on the output. As usual, we can try this behavior by running the server and making a request using `curl`.

The onError hook

This hook is triggered only when the server sends an error as the payload to the client. It runs after `customErrorHandler` if provided or after the default one integrated into Fastify. Its primary use is to do additional logging or to modify the reply headers. We should keep the error intact and avoid calling `reply.send` directly. The latter will result in the same error we encountered trying to do the same inside the `onResponse` hook. The snippet shown in the `on-error.cjs` example makes it easier to understand:

```
const Fastify = require('fastify')

const app = Fastify({ logger: true })
app.addHook('onError', async (request, _reply, error) => {
// [1]
  request.log.info(`Hi from onError hook:
    ${error.message}`)
})
app.get('/foo', async (_request, _reply) => {
  return new Error('foo') // [2]
})
app.get('/bar', async (_request, _reply) => {
  throw new Error('bar') // [3]
})

app.listen({ port: 3000 }).catch(err => {
  app.log.error(err)
  process.exit()
})
```

First, we define an `onError` hook at **[1]** that logs the incoming error message. We don't want to change the `error` object to return any value from this hook, as we already said. So then, we define two routes: `/foo` (**[2]**) returns an error while `/bar` (**[3]**) throws an error.

We can run the snippet in a terminal:

```
$ node on-error.cjs
{"level":30,"time":1636719503620,"pid":62791,"hostname":"localhost",
"msg":"Server listening at http://127.0.0.1:3000"}
```

Now, in a different terminal window, we can make two different calls to our server:

```
$ curl localhost:3000/bar
{"statusCode":500,"error":"Internal Server Error","message":"bar"}
> curl localhost:3000/foo
{"statusCode":500,"error":"Internal Server Error","message":"foo"}
```

Checking the server log terminal will show us that, in both cases, the onError hook was triggered correctly.

The onTimeout hook

This is the last hook we still need to discuss. It depends on the connectionTimeout option, whose default value is 0. Therefore, onTimeout will be called if we pass a custom connectionTimeout value to the Fastify factory. In on-timeout.cjs, we use this hook to monitor the requests that time out. Since it is only executed when the connection socket is hung up, we can't send any data to the client:

```
const Fastify = require('fastify')
const { promisify } = require('util')

const wait = promisify(setTimeout)
const app = Fastify({ logger: true, connectionTimeout: 1000 }) // [1]
app.addHook('onTimeout', async (request, reply) => {
  request.log.info(`The connection is closed.`) // [2]
})

app.get('/', async (_request, _reply) => {
  await wait(5000) // [3]
  return ''
})

app.listen({ port: 3000 }).catch(err => {
  app.log.error(err)
  process.exit()
})
```

At **[1]**, we pass the connectionTimeout option to the Fastify factory, setting its value to 1 second. Then we add an onTimeout hook that prints to logs every time a connection is closed (**[2]**). Finally, we add a route that waits for 5 seconds before responding to the client (**[3]**).

Let's run the snippet:

```
$ node on-timeout.cjs
{"level":30,"time":1636721785344,"pid":63255,"hostname":"localhost",
"msg":"Server listening at http://127.0.0.1:3000"}
```

Now from a different terminal window, we can make our call:

```
$ curl localhost:3000
curl: (52) Empty reply from server
```

The connection was closed by the server, and the client received an empty response.

We can return to the application log terminal window and check the output to see that our `onTimeout` hook was called:

```
{"level":30,"time":1636721844526,"pid":63298,"hostname":"localhost","
reqId":"req-1","req":{"method":"GET","url":"/","hostname":"localhost:
3000","remoteAddress":"127.0.0.1","remotePort":65021},"msg":"incoming
request"}
{"level":30,"time":1636721845536,"pid":63298,"hostname":"
localhost","reqId":"req-1","msg":"The connection is closed."}
```

Replying from a hook

Besides throwing an error, there is another way of early exiting from the request phase execution flow at any point. Terminating the chain is just a matter of sending the reply from a hook: this will prevent everything that comes after the current hook from being executed. For example, this can be useful when implementing authentication and authorization logic.

However, there are some quirks when dealing with asynchronous hooks. For example, after calling `reply.send` to respond from a hook, we need to return the `reply` object to signal that we are replying from the current hook.

The `reply-from-hook.cjs` example will make everything clear:

```
app.addHook('preParsing', async (request, reply) => {
  const authorized = await isAuthorized(request) // [1]
  if (!authorized) {
    reply.code(401)
    reply.send('Unauthorized') //[2]
    return reply // [3]
  }
})
```

We check whether the current user is authorized to access the resource **[1]**. Then, when the user misses the correct permissions, we reply from the hook directly **[2]** and return the `reply` object to signal it **[3]**. We are sure that the hook chain will stop its execution here, and the user will receive the `'Unauthorized'` message.

Route-level hooks

Until now, we declared our hooks at the application level. However, as we mentioned previously in this chapter, request/response hooks can also be declared on a route level. Thus, as we can see in `route-level-hooks.cjs`, we can use the route definition to add as many hooks as we like, allowing us to perform actions only for a specific route:

```
app.route({
  method: 'GET',
```

```
    url: '/',
    onRequest: async (request, reply) => {},
    onResponse: async (request, reply) => {},
    preParsing: async (request, reply) => {},
    preValidation: async (request, reply) => {},
    preHandler: async (request, reply) => {},
    preSerialization: async (request, reply, payload) => {},
    onSend: [async (request, reply, payload) => {}, async
      (request, reply, payload) => {}], // [1]
    onError: async (request, reply, error) => {},
    onTimeout: async (request, reply) => {},
    handler: function (request, reply) {},
  })
```

If we need to declare more than one hook for each category, we can define an array of hooks instead ([1]). As a final note, it is worth mentioning that these hooks are always executed as last in the chain.

Route-level hooks conclude this section. It is undoubtedly a longer one, but it is packed with concepts that are the core of the Fastify framework.

Summary

In this chapter, we learned one of the Fastify concepts that make it different from most web frameworks. Even if it might be a new concept to many developers, mastering hooks will unleash the path to highly performant, maintainable, and reusable application components. In fact, contrary to middleware, when we define a hook, we are guaranteed that our code is executed only when it makes sense, boosting the application's performance. Furthermore, having a well-defined order of execution can help with debugging and runtime checking.

In the next chapter, we will learn how to make user inputs safe and speed up our server thanks to validation and serialization.

5

Exploring Validation and Serialization

Fastify is secure and fast, but that doesn't protect it from misuse. This chapter will teach you how to implement secure endpoints with input validation and make them faster with the serialization process.

This framework provides all the tools you need to take advantage of these two critical steps, which will support you while exposing straightforward API interfaces and enable your clients to consume them.

You will learn how to use and configure Fastify's components in order to control and adapt the default setup to your application logic.

This is the learning path we will cover in this chapter:

- Understanding validation and serialization
- Understanding the validation process
- Customizing the validator compiler
- Managing the validator compiler
- Understanding the serialization process

Technical requirements

As mentioned in the previous chapters, you will need the following:

- A working Node.js 18 installation
- A text editor to try the example code
- An HTTP client to test out code, such as CURL or Postman

All the snippets in this chapter are on GitHub at `https://github.com/PacktPublishing/ Accelerating-Server-Side-Development-with-Fastify/tree/main/Chapter%205`.

Understanding validation and serialization

Fastify has been built with a focus on the developer's experience, and on reducing the developer effort needed to draft a new project. For this reason, Fastify has built-in features to reduce the following burdens:

- Validating the user's input
- Filtering the server's output

The aim is to find solutions for and prevent the most common security attacks, such as code injection or sensitive data exposure. The answer is declaring the expected input and output data format for every route. Therefore, the validation and serialization processes have been introduced into the framework by design:

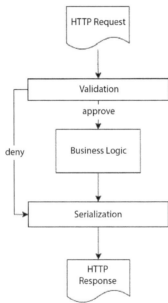

Figure 5.1 – The Validation and Serialization phases

This preceding diagram shows the request lifecycle steps' macro architecture, which you read about in detail in *Chapter 4*.

The **Validation** phase happens when the **HTTP Request** comes into the server. It allows you to approve or deny access to the **Business Logic** step.

The **Serialization** step transforms high-level data produced by the business logic, such as JSON objects or errors, into low-level data (`strings` or `buffers`) to reply to the client's request.

The next question is: how do you define the information passing through the validation and the response data? Fortunately, the solution is the **JSON Schema** specification, which is embraced by both Fastify and the web community.

But what is JSON Schema? We are going to understand this important application's concept, which focuses on security and speed. The following sections are more theoretical than practical: we need to know how the system works before seeing it in action, otherwise, we may miss important concepts.

The JSON Schema specification

The JSON Schema standard describes the structure of JSON documents. Therefore, by using a JSON Schema interpreter, it is possible to verify whether a JSON object adapts to a defined structure and act accordingly.

Writing a schema gives you the possibility to apply some automation to your Node.js application:

- Validating JSON objects
- Generating documentation
- Filtering JSON object fields

The standard is still in the draft phase, and it has reached version 2020-12. By default, Fastify v4 adopts an older version of the specification, **Draft-07**, which is broadly supported and used. For this reason, all the next JSON Schema examples follow this standard. Let's see in practice what a JSON Schema looks like by trying to validate the following JSON object through a schema:

```
{
  "id": 1,
  "name": "Foo",
  "hobbies": [ "Soccer", "Scuba" ]
}
```

The corresponding JSON Schema could assume the following structure:

```
{
  "$schema": "http://json-schema.org/draft-07/schema#",
  "$id": "http://foo/user",
  "type": "object",
  "properties": {
    "identifier": {
      "type": "integer"
    },
    "name": {
```

```
      "type": "string",
      "maxLength": 50
    },
    "hobbies": {
      "type": "array",
      "items": {
        "type": "string"
      }
    }
  },
  "required": ["identifier", "name"]
}
```

This schema validates the initial JSON object. As you can see, the schema is JSON that is readable even without knowing the standard format. Let's try to understand it together.

We expect a `type` object as input with some properties that we have named and configured as follows:

- The required `identifier` field must be an integer
- A mandatory `name` string that cannot be longer than 50 characters
- An optional `hobbies` string array

In this case, a software interpreter's output that checks if the input JSON is compliant with the schema would be successful. The same check would fail if the input object doesn't contain one of the mandatory fields, or if one of the types doesn't match the schema's `type` field.

So far, we have talked about the JSON object's validation, but we haven't mentioned the serialization. These two aspects are different, and they share the JSON Schema specification only. The specification is written keeping the validation process in mind. The serialization is a nice "side effect," introduced to improve security and performance; we will see how within this section.

The example schema we have just seen is a demo showing the basic schema syntax, which will become more intuitive. JSON Schema supports a large set of keywords to implement strict validation, such as default values, recursive objects, date and time input types, email format, and so on.

Providing a complete overview of the JSON Schema specification could take up the whole book. This is why you can deepen your knowledge of this aspect by checking the official website at `https://json-schema.org/`. We will have looked at other keywords by the end of this chapter and new ones will be introduced and described gradually.

As you may have noticed reading the previous schema example, some keywords have the dollar symbol prefix, `$`. This is special metadata defined in the draft standard. One of the most important and most used ones is the `$id` property. It identifies the JSON Schema univocally, and Fastify relies upon it to process the schema objects and reuse them across the application.

The $schema keyword in the example tells us the JSON document's format, which is Draft-07. Whenever you see a JSON Schema in these pages, it is implicit that it follows that version due to Fastify's default setup.

Now we have an idea of what a schema is, but how does it integrate with Fastify's logic? Let's find out.

Compiling a JSON Schema

A JSON Schema is not enough to validate a JSON document. We need to transform the schema into a function that our software can execute. For this reason, it is necessary to use a **compiler** that does the work.

It is essential to understand that a JSON Schema compiler tries to implement the specification, adding valuable features to ease our daily job. This implies knowing which compiler your application uses to tweak the configuration and how to benefit from some extra features such as new non-standard keywords, type coercion, and additional input formats.

> **Compiler implementation lock-in**
>
> Generally, writing JSON Schema by using a new compiler's keywords and features leads to lock-in. In this case, you will not be able to change the compiler, and you may find issues during integrations that rely on standard JSON schemas only, such as API document generation. This is fine if you consider the pros and cons that we will present in this chapter.

The same logic has been carried out in the serialization process. The idea was quite simple: if it is possible to build a JavaScript function to validate a JSON object, it is possible to compile a new function that produces a string output. The string would be based only on the fields defined in the JSON schema's source!

By following this step, you can define only the data you want to enter your server and go out of the application! This improves the application's security. In fact, the compiler's implementation has a secure mechanism to block code injection, and you can configure it to reject bad input data, such as strings that are too long.

We have now clarified the JSON Schema and explained how it can help improve an application within a compiler component. Let's understand how Fastify has integrated this logic into the framework.

Fastify's compilers

Fastify has two compilers by default:

- **The Validator Compiler**: Compiles the JSON Schema to validate the request's input
- **The Serializer Compiler**: Compiles the response's JSON Schema to serialize the application's data

These compilers are basically Node.js modules that take JSON Schema as input and give us back a function. Keep this in mind because it will be important later, in the *Building a new validator compiler* section.

Fastify's detailed workflow can be schematized as follows:

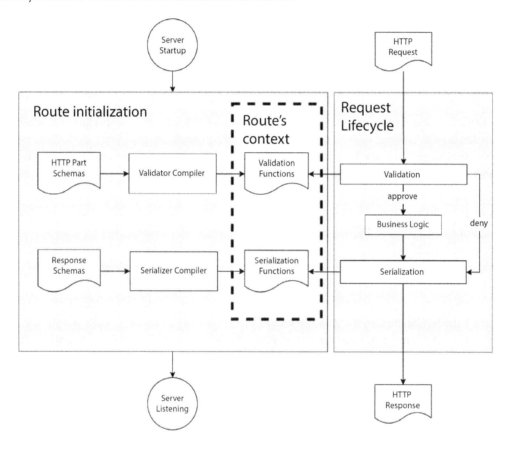

Figure 5.2 – Fastify's JSON Schema compilation workflow

As you can see, there are two distinct processes:

- The **Route initialization**, where the schemas are compiled during the startup phase.

- The request through the **Request Lifecycle**, which uses the compiled functions stored in the route's context.

Now you should have a complete overview of Fastify's generic components and the logic they implement. It's time to see it all in action. Note that to ease understanding and avoid confusion, we will discuss validation and serialization separately.

Understanding the validation process

The validation process in Fastify follows the same logic to validate the incoming HTTP request parts. This business logic comprises two main steps, as we saw in *Figure 5.2*:

- The schema compilation executed by the **Validator Compiler**
- The validation execution

We will discuss these aspects one by one.

The validator compiler

Fastify doesn't implement a JSON Schema interpreter itself. Still, it has integrated the **Ajv** (`https://www.npmjs.com/package/ajv`) module to accomplish the validation process. The Ajv integration into Fastify is implemented to keep it as fast as possible and support the encapsulation as well. You will always be able to change the default settings and provide a new JSON Schema interpreter to the application, but we will learn how to do it later, in the *Managing the validator compiler* section.

> **Ajv version**
>
> Fastify has included the Ajv module version 8. This is important to know when you need to search for new configurations or to ask for support from the Ajv community. Moreover, the Ajv version defines the supported JSON Schema versions. At the time of writing, the lastest Ajv module version is v8, which supports the 2020-12 specification.

The validator compiler component is a factory function that must compile the application's route schemas. Every route may define one schema per HTTP part:

- `params` for the path parameters
- `body` for the payload
- `querystring` (or the `query` alias field) for the URL's query strings
- `headers` for the request's headers

All these properties are optional, so you can choose freely which HTTP part has to be validated.

The schemas must be provided during the route declaration:

```
app.post('/echo/:myInteger', {
  schema: {
    params: jsonSchemaPathParams,
    body: jsonSchemaBody,
    querystring: jsonSchemaQuery,
    headers: jsonSchemaHeaders
```

```
    }
}, function handler (request, reply) {
    reply.send(request.body)
})
```

You are done! Now, whenever you start the application, the schemas will be compiled by the default validator compiler. The generated functions will be stored in the route's context, so every HTTP request that hits the /echo/:myinteger endpoint will execute the validation process.

We can call our endpoint with the incorrect data to check our code, for example, /echo/not-a-number. This input will trigger a validation error, and we will get back a 400 Bad Request response:

```
{
    "statusCode":400,
    "error":"Bad Request",
    "message":"params.myInteger should be integer"
}
```

As we saw, the compilation seems relatively easy, but you also need to know that this feature is *fully encapsulated*. This architectural pattern, which we have discussed already, is designed as follows:

- One validation compiler per different plugin context and they will not collide
- You can add schemas with the same $id in different contexts and they may have different structures

Before further discussing the validator compiler and how to configure and change it, let's continue on this "happy path" to get a complete picture of one of Fastify's key aspects.

Validation execution

Fastify applies the HTTP request part's validation during the request lifecycle: after executing the preValidation hooks and before the preHandler hooks.

The purpose of this validation is to check the input format and to produce one of these actions:

- **Pass**: Validates the HTTP request part successfully
- **Deny**: Throws an error when an HTTP request part's validation fails
- **Append error**: When an HTTP request part's validation fails and continues the process successfully – configuring the attachValidation route option

This process is not designed to verify data correctness – for that, you should rely on the preHandler hook.

As you have seen in the preceding code example, the schema object has a defined structure, where every property maps an HTTP part: params, body, querystring, and headers. When you set a JSON Schema to body, the HTTP request payload must be JSON input by default. You can overwrite this behavior, and we will see how to do it in the next section.

In the previous chapters, all of our route examples did not have the `schema` route option. By doing so, *we skipped the validation* phase of the request lifecycle.

> **Validation execution order**
>
> The HTTP part list mentioned in *The validator compiler* section is ordered by execution. This means that if the `params` validation fails, the subsequent HTTP parts will not be processed.

The validation process is quite straightforward. Let's zoom in on this process' logic, looking at the entire request lifecycle:

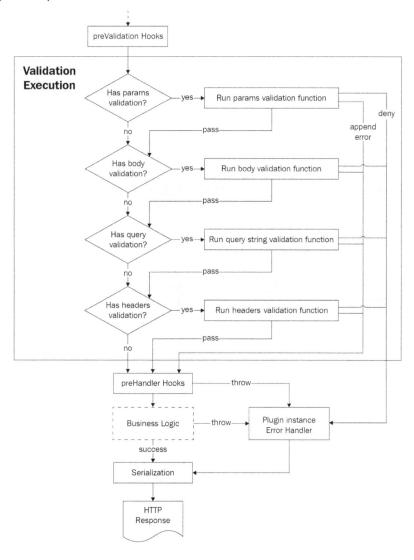

Figure 5.3 – The validation execution workflow

Let's understand the flow diagram step by step:

- The dotted arrow is an HTTP request that has started its lifecycle into the Fastify server and has reached the `preValidation` hooks step. All will work as expected, and we are ready to start the **Validation Execution**.

- Every HTTP part is validated if a JSON Schema has been provided during the route's declaration.

- The validation passes and proceeds to the next step.

- When the validation fails, a particular `Error` object is thrown, and it will be processed by the **error handler** configured in the server instance where the route has been registered. Note that the error is suppressed if the `attachValidation` route option is set. We will look at an example in the *Flow control* section.

- If all the validations are successful, the lifecycle continues its flow to the `preHandler` hooks, and it will continue as discussed in the previous chapters.

- The **Business Logic** dashed box represents the handler execution that has been omitted because the image is specifically focused on validating the execution flow.

These steps happen when the `schema` option is set into the route definition, as in the previous code snippet in *The validator compiler* section.

Now we have a complete overview of the entire validation process, from the startup to the server's runtime. The information provided covers the most common use cases for an application and, thanks to Fastify's default settings, it is ready to use.

Great applications need great features. This is why we will now focus on the validator compiler customization.

Customizing the validator compiler

Fastify exposes a lot of options to provide a flexible validation process and complete control of it. We are going to explore all the possible customizations one by one, so you will be a validator compiler guru by the end of this section! Let's jump into this journey one step at a time!

Flow control

In the previous section, we mentioned the `attachValidation` route option – it's time to look at an example (although you probably already know how to use it, thanks to the previous chapters):

```
app.get('/attach-validation', {
  attachValidation: true,
  schema: {
    headers: jsonSchemaHeaders
  },
```

```
    handler: (request, reply) => {
      reply.send(request.validationError)
    }
  })
```

Adding the flag into the route option will prevent a validation error from being thrown. Instead, the validation execution process will be interrupted at the first error occurrence, and the process will continue as the validation has been successful. In this case, a `validationError` object will be attached to the `request` argument. The subsequent route's entities in the request lifecycle have to deal with the error and act accordingly or the error will not be managed. As in the previous code example, the handler function is always executed.

Understanding the Ajv configuration

The Ajv configuration defines how the validation's functions are built and how they will behave in some circumstances. The default settings are the following, and it is worth knowing about them:

```
{
  coerceTypes:'array',
  useDefaults: true,
  removeAdditional: true,
  uriResolver: require('fast-uri'),
  allErrors: false,
  nullable: true
}
```

Let's get an understanding of them, and then we will provide an example to see all these options in action:

- The `coerceTypes` flag tries to coerce the input data to the type defined in the schema. For example, if an input body property is the string `foo: "42"`, and if the field itself is defined as `type: integer`, the `request.body.foo` field will be coerced to `Number`. We will investigate the `array` value later in this section.

- The `useDefaults` option will enable the use of the `default` JSON Schema keyword, letting you define an initial value if a property is missing or undefined.

- The `removeAdditional` setting allows you to evict all the properties that are not listed in the JSON Schema from the HTTP part field.

- `uriResolver` is a parameter introduced by the Fastify community. It speeds up the Ajv module processing even more.

- A JSON object may have multiple validation errors, such as two fields that are not the correct data type. The `allErrors` flag configures the validation function to stop at the first error occurrence.

- The `nullable` flag lets you use the `nullable` keyword's syntactic sugar in your JSON schemas.

These options and more are well documented on the Ajv site at `https://ajv.js.org/options.html`. You can refer to them to find new options or to change the default ones. We will look at a couple of the most used configurations as a baseline in the *Configuring the default Ajv validator compiler* section.

It is important to note that these options let the validation function manipulate the original request's input. This implies that the raw body is processed and modified.

How is the preValidation hook born?

The `preValidation` hook was first introduced in Fastify's core due to the raw body manipulation that the validation functions were doing. This has been necessary in limited cases only, such as a signed body payload that requires an unmodified client's input.

To see it all in action, here is a JSON Schema:

```
const ajvConfigDemoSchema = {
  type: 'object',
  properties: {
    coerceTypesDemo: { type: 'integer' },
    useDefaultsDemo: { type: 'string', default: 'hello' },
    removeAdditional: {
      type: 'object',
      additionalProperties: false,
      properties: {
        onlyThisField: {
          type: 'boolean'
        }
      }
    },
    nullableDemo: { type: 'string', nullable: true },
    notNullableDemo: { type: 'string' }
  }
}
```

This schema introduces three new keywords:

- The `default` property lets you define a default value when the JSON input object does not contain the `useDefaultsDemo` property or its value is `null`.

- The `additionalProperties` parameter is used to control the handling of extra properties. In the example, you see the `boolean false`, which evicts the additional data from the HTTP part. An object could also apply more complex filters. Please refer to the official specification: `https://json-schema.org/understanding-json-schema/reference/object.html#additional-properties`.

- The `nullable` flag is not defined in the standard. It is syntactic sugar to avoid the standard type definition for `nullable` fields: `{ type: ["string", "null"] }`.

Using this schema in a route handler will give us a clear understanding of the configured options:

```
app.post('/config-in-action', {
  schema: {
    body: ajvConfigDemoSchema
  },
  handler (request, reply) {
    reply.send(request.body)
  }
})
```

Calling the endpoint, defined with the following payload, should set a reply with the modified body after the validation function's execution:

```
curl --location --request POST 'http://localhost:8080/config-in-
action' \
--header 'Content-Type: application/json' \
--data-raw '{
    "coerceTypesDemo": "42",
    "removeAdditional": {
        "remove": "me",
        "onlyThisField": true
    },
    "nullableDemo": null,
    "notNullableDemo": null
}'
```

We should get back this response as an output:

```
{
    "coerceTypesDemo": 42,
    "removeAdditional": {
        "onlyThisField": true
    },
    "nullableDemo": null,
    "notNullableDemo": "",
    "useDefaultsDemo": "hello"
}
```

The changes have been highlighted, and each property name describes the Ajv option that triggered the change.

So far, we have a complete understanding of the default validator compiler's configuration. This covers the most common use cases and gives you the possibility to use it out of the box without struggling with complex configuration or having to learn about the `Ajv` module. Unfortunately, it is crucial to control all Fastify's components and configuration in order to manage a real-world application. In the following section, you will learn how to customize the validator compiler.

Managing the validator compiler

Fastify offers you the possibility to customize the validator compiler in two different manners:

- Configuring the default Ajv validator compiler
- Implementing a brand-new validator compiler, such as a new JSON Schema compiler module

These options give you total control over the validation process and the ability to react to every situation you may face, such as adopting a new validator compiler module or managing how the Ajv package processes the input data.

Configuring the default Ajv validator compiler

In the *Understanding the Ajv configuration* section, we saw the default Ajv settings and a link to its documentation to explore them all. If you find some useful options you would like to apply, you can set them during the Fastify instance instantiation:

```
const app = fastify({
  ajv: {
    customOptions: {
      coerceTypes: 'array',
      removeAdditional: 'all'
    },
    plugins: [
      [
        require('ajv-keywords'),
        'transform'
      ]
    ]
  }
})
```

Fastify's factory accepts an Ajv option parameter. The parameter has two main fields:

- `customOptions` lets you extend Ajv's settings. Note that this JSON will be merged within the default settings.
- The `plugins` array accepts the Ajv's external plugins.

The new settings used in the example are the ones I prefer the most. The `coerceTypes` value solves the issue when you need to receive an array parameter via `querystring`:

```
app.get('/search', {
  handler: echo,
  schema: {
    query: {
      item: {
        type: 'array',
        maxItems: 10
      }
    }
  }
})
```

Without the `coerceTypes: 'array'`, if your endpoint receives just one parameter, it won't be coerced to an array within one element by default, thus leading to an error of type mismatch. Note that this option is already set as a default by Fastify.

The `removeAdditional` option value makes it possible to avoid redefining `additional Properties: false` in all our schema objects. Note that it is crucial to list all the properties in the application's schemas, or you will not be able to read the input in your handlers!

JSON Schema shorthand declaration

In the previous example, the query's schema didn't have some of the mandatory JSON Schema fields: `type` and `properties`. Fastify will wrap the input JSON Schema in parent JSON Schema scaffolding if it does not recognize the two properties. This is how Fastify's syntactic sugar works, to ease the route's configuration.

After the `customOptions` Ajv configuration option field, it is possible to set the `plugins` property. It adds new features and keywords to the JSON Schema specification, improving your developer experience.

The `plugins` option must be an array, where each element should be either of the following:

- The Ajv plugin's function
- A two-element array, where the first item is the Ajv plugin's function and the second is the plugin's options

We can see how to use it in the following snippet. We are registering the same plugin multiple times for the sake of showing the syntaxes:

```
plugins: [
  require('ajv-keywords'), // [1]
```

```
[ // [2]
  require('ajv-keywords'),
  'transform'
]
```

As you have seen, Fastify's validator compiler is highly configurable and lets you find the best settings for your application. We have almost covered all the settings that Fastify exposes, in order to configure the default compiler.

So far we have used the validation output as is, but if you are asking yourself whether it is customizable, of course it is! Let's see how to do it.

The validation error

The validator function is going to throw an error whenever an HTTP part doesn't match the route's schema. The route's context error handler manages the error. Here is a quick example to show how a custom error handler could manage an input validation error in a different way:

```
app.get('/custom-error-handler', {
  handler: echo,
  schema: {
    query: { myId: { type: 'integer' } }
  }
})
app.setErrorHandler(function (error, request, reply) {
  if (error.validation) {
    const { validation, validationContext } = error
    this.log.warn({ validationError: validation })
    const errorMessage = `Validation error on
    ${validationContext}`
    reply.status(400).send({ fail: errorMessage })
  } else {
    this.log.error(error)
    reply.status(500).send(error)
  }
})
```

As you can see, when the validation fails, two parameters are appended to the Error object:

• The validationContext property is the HTTP part's string representation, responsible for generating the error

• The validation field is the raw error object, returned by the validator compiler implementation

The default Fastify error handler manages the Ajv error object and returns a clear error message.

The validation error data type

The default compiler produces an Ajv error array. Therefore, the `validation` property is generated by Ajv's compiled function. Whenever we use the custom validator compiler with a new error format, the `validation` field mutates its data type accordingly.

Customizing the error handler gives you the control to make the validation errors conform to your application's error format output. We saw an example earlier, in the *The validator compiler* section.

If you just need to customize the error message instead, Fastify has an option even for that! The `schemaErrorFormatter` option accepts a function that must generate the `Error` object, which will be thrown during the validation process flow. This option can be set in the following ways:

- During the root server initialization
- As the route's option
- Or on the plugin registration's instance

Here is a complete overview of the three possibilities in the same order as in the preceding list:

```
const app = fastify({
  schemaErrorFormatter: function (errors, httpPart) { //[1]
    return new Error('root error formatter')
  }
})
app.get('/custom-route-error-formatter', {
  handler: echo,
  schema: { query: { myId: { type: 'integer' } } },
  schemaErrorFormatter: function (errors, httpPart) { //[2]
    return new Error('route error formatter')
  }
})
app.register(function plugin (instance, opts, next) {
  instance.get('/custom-error-formatter', routeConfig)
  instance.setSchemaErrorFormatter(function (errors,
    httpPart) { // [3]
    return new Error('plugin error formatter')
  })
  next()
})
```

The `setSchemaErrorFormatter` input function must be synchronous. It is going to receive the raw `errors` object returned by the compiled validation function, plus the part of HTTP that is not valid.

So far, we have tweaked the default Fastify validator compiler, since it generates the validation function for the error output. There are quite a lot of settings, but they allow you to customize your server based on your choices, without dealing with the compilation complexity. We still have to explain how to change the validator compiler implementation, but we must learn how to reuse JSON schemas first.

Reusing JSON schemas

JSON schemas may seem huge and long to read and understand at first sight. In fact, in the *The JSON Schema specification* section, we saw a ~20-line schema to validate a three-field JSON object.

The JSON Schema specification solves this issue by providing schema reusability through the `$ref` keyword. This property is used to reference a schema and must be a string URI. `$ref` may reference an external JSON Schema or a local one in the schema itself.

To reference an external schema, we must start with the following two actions:

1. Set the `$id` property of the external schema and the `$ref` values to reference it.

2. Add the external schema to the Fastify context, calling the `app.addSchema(json Schema)` method.

To understand it better, we are going to use an example:

```
app.addSchema({
  $id: 'http://myapp.com/user.json',
  definitions: {
    user: {
      $id: '#usermodel',
      type: 'object',
      properties: {
        name: { type: 'string', maxLength: 50 }
      }
    },
    address: {
      $id: 'address.json',
      definitions: {
        home: { $id: '#house', type: 'string', maxLength:
        150 },
        work: { $id: '#job', type: 'string', maxLength: 200
        }
      }
    }
  }
```

```
    }
})
```

The `addSchema` method accepts a valid JSON Schema as an argument, and it must have the `$id` value. Otherwise, the schema can't be referenced. If the `$id` value is missing, an error is thrown. Adding a schema by following this example lets us reference it in the route configuration's `schema` property.

The schema's $id

In the previous code block, the `$id` value is an absolute **Uniform Resource (URI)**. The JSON Schema specification defines that the root schema's `$id` should be in this format. The URI set doesn't need to be a real HTTP endpoint as in the example. It must be unique. Following the specification will help you adopt external tools to manipulate your application's schemas, such as documentation generation. As an example, I like to use a URI in this format: `schema:myapplication:user:create`, which can be summarized as `schema:<application code>:<model>:<scope>`.

To reference the `http://myapp.com/user.json` schema, we must use the `$ref` keyword:

```
app.post('/schema-ref', {
  handler: echo,
  schema: {
    body: {
      type: 'object',
      properties: {
        user: { $ref:
          'http://myapp.com/user.json#usermodel' }, // [1]
        homeAdr: { $ref:
          'http://myapp.com/address.json#house' }, // [2]
        jobAdr: { $ref:
          'http://myapp.com/address.json#/definitions/work'
          // [3]
        },
        notes: { $ref: '#/definitions/local' } // [4]
      },
      definitions: {
        local: { type: 'boolean' }
      }
    }
  }
})
```

We used four different URI reference formats. Generally, the $ref format has this syntax:

```
<absolute URI>#<local fragment>
```

Here is a brief explanation:

1. **Reference to an external fragment**: The user's property points to an external $id, defined in the user.json URI domain.

2. **Reference to an external subschema fragment**: homeAdr has replaced the absolute URI, from user.json to address.json. This happens because if $id doesn't start with the # char, it is a relative path to the root URI. So, in the external schema, we have defined an address.json subschema.

3. **Relative path to an external subschema**: The local fragment can be a relative path to the JSON Schema to apply. Note that the relative path uses the JSON field name work and not the $id value.

4. **Local reference**: You can reference the schema itself. When $ref does not have an absolute URI before the # char, the local fragment is resolved locally.

This example gives you a complete tool belt and helps you to define your own schema's references. This setup covers the most common use cases. In fact, the specification is comprehensive and covers many more use cases that could not adapt to our applications. You may refer to the official documentation example to deep dive into $ref and $id linking https://datatracker.ietf.org/doc/html/draft-handrews-json-schema-01#section-8.2.4.

We have learned how to share schemas across the application and use them in our routes' configuration. But how can we read them? Let's find out in the next section.

Retrieving your schemas

You must be aware that the schemas added through the addSchema() method are stored in a bucket component. Every plugin instance has one bucket instance to support the encapsulation. Let's see this code:

```
app.register(async (instance, opts) => {
  instance.addSchema({
    $id: 'http://myapp.com/user.json',
    type: 'string',
    maxLength: 10
  })
})
app.register(async (instance, opts) => {
  instance.addSchema({
    $id: 'http://myapp.com/user.json',
    type: 'string',
```

```
      maxLength: 50
    })
  })
```

Adding two schemas with the same $id is going to throw an error during the startup phase. The usual Fastify logic will apply as always: schemas added in the plugin's parent scope are inherited in the child one. But, two schemas can have the same $id only in different encapsulated contexts.

> **Fastify loves optimizing**
>
> As mentioned, every plugin context has a `bucket` component that contains all the context's schemas. This is not totally true. The schema objects could be vast and repetitive. Therefore, Fastify optimizes the bucket logic to reduce the `bucket` instances to the minimum that is needed to isolate them in an encapsulated context. For example, adding all the shared application's schemas in the root context leads to a single `bucket` instance.

To read the application's schemas from the `bucket` component, you can use these methods:

```
const json =
instance.getSchema('http://myapp.com/user.json')
const jsonIdSchemaPair = instance.getSchemas()
```

`getSchema(id)` needs a string argument within the $id to retrieve the corresponding schema. The `getSchemas()` method returns a JSON key-value pair where the key is the schema's $id and the value is the JSON Schema itself.

Now we know about all the aspects of the default validator compiler, how can we configure and extend it? It is time to drop it in favor of a custom one!

Building a new validator compiler

Fastify's validator compiler can be replaced totally. You may need to do this in the following situations:

- You need a unique validator compiler for each HTTP part. For example, you need a specific Ajv configuration for the `body` part and a different configuration for `querystring`.

- You prefer to use a different JSON Schema compiler module, such as `@hapi/joi` or yup.

- You want to build your compiler to optimize your application.

It is important to keep in mind that rewriting this Fastify component requires knowing the validation flow, because not being able to find the schemas is one of the most common pitfalls.

Fastify provides you with a couple of ways to substitute the compiler:

- The root instance's **Schema Controller Factory**
- The root or plugin's validator or serializer compiler component

These two options interact with each other as shown in the following diagram:

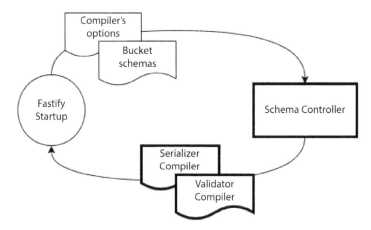

Figure 5.4 – The compiler's customization process

During the startup, Fastify decides when it needs a validator compiler or serializer compiler, based on the following questions:

- Does the root or the plugin instance need a compiler?
- Does the root or the plugin instance have a compiler already?
- Are the `bucket` schemas changed, compared to the parent instance?

When all these conditions are met, Fastify asks the schema controller factory to build a new compiler. Note that the following logic applies to both the validator compiler and the serializer compiler. We are going to focus on the validator compiler, but the flow is the same for each of them. We will explore the serializer compiler in the *The serializer compiler* section.

The input to produce the compiler is as follows:

- The compiler option. For the validation compiler, it corresponds to the application's `ajv` option property.
- The instance's `bucket` schemas.

Within these arguments, the factory must produce a Fastify compiler.

The validator compiler is a function that, given the route's schemas, must return a new function to validate the HTTP request's parts – we know, it seems a bit tricky, but let's look at a complete example to understand this idea:

```
const app = fastify({
  schemaController: { // [1]
    compilersFactory: {
      buildValidator: myCompilerFactory // [2]
    }
  }
})
app.addSchema({
  $id: 'http://myapp.com/string.json',
  type: 'string',
  maxLength: 50
})
function myCompilerFactory (externalSchemas,
ajvServerOption) {
  return myValidatorCompiler // [3]
}
function myValidatorCompiler (routeData) {
  const { schema, method, url, httpPart } = routeData
  return function validate (jsonPayload) { // [4]
    return true
  }
}
```

The example shows all the schema controller's entities, which are just simple functions that don't really do anything. Still, they provide us with an excellent overview of how the single parts can work together, adding some `console.log` statements during your experiments.

> **Disclaimer**
>
> Be patient and try to follow the code: we are going to explore a function that returns a function that returns another function! It is pretty easy to get lost when first reading it, but hang in there!

There is a new `schemaController` option [1]. This parameter lets us configure and take control of the compilers through the `compilersFactory` object. It accepts two parameters, `buildValidator` and `buildSerializer`.

We are going to focus on the former: providing a function [2] as input. The `myCompilerFactory` function can access all the instance's schema via the `externalSchemas` argument and `ajvServerOption`. In this example, there is just one external schema, `'http://myapp.com/string.json'`, and the ajv server options are empty.

> **External schemas**
>
> The `externalSchemas` argument is provided by Fastify's schemas bucket. It is populated by calling `instance.getSchemas()` internally. The word "external" refers to those schemas that are referenced by the routes' schemas.

`myCompilerFactory` is executed whenever Fastify needs a new compiler, based on the checklist we saw at the beginning of this section. For example, the smallest memory footprint case is set by calling this function once: there is only one validator compiler for the whole application.

> **Fastify uses the same APIs exposed to the user**
>
> It is important to mention that Fastify uses this API to implement the default schema controller. The default `myCompilerFactory` implementation creates the default Ajv instance, and it is isolated in an external module named `@fastify/ajv-compiler`. Thanks to this module, it is possible to run Fastify with a different Ajv version out of the box. Please give it a check!

The `myCompilerFactory` function returns the `myValidatorCompiler` function [**3**]. The latter function target is to "compile" the input schema and transform it into an executable JavaScript function. The `myValidatorCompiler` function is executed during the startup phase, once for every route's JSON Schema. The `routeData` argument is an object within the route's coordinates where the following applies:

* `schema` is the object provided to the route's `schema` option (sorry for the redundancy).
* `method` is the route's HTTP method string in uppercase.
* `url` is the raw route's URL string. For example, the path parameters such as `/hello/:name`.
* `httpPart` tells us which HTTP request's part should be validated. It can be one of the well-known `body`, `params`, `querystring`, or `headers` strings.

The `myValidatorCompiler` [**3**] function returns... another function! It is the last one, I swear. The `validate` [**4**] function is what the compilation of the schema must produce. It is stored in the route's context and executed for every HTTP request routed to that endpoint. We saw the schema of this process in *Figure 5.3*.

The `validate` function is the one that is run when the server is listening. The `schemaController` flow, we have just seen, is executed once during the startup, and it is the heaviest task Fastify must complete before accepting incoming requests. To put it in practice, Fastify's default `schemaController` component uses the Ajv module to build these `validate` functions – no more, no less.

In the example, we customized `schemaController` with a validation that always returns `true`. But we could implement our logic based on `httpPart`, such as using a primary validation function for `querystring` and `headers` and a more complex one for the `body` part. The most common use case is to apply different ajv settings for each HTTP part. You can find a complete example at `https://github.com/fastify/help/issues/128`.

Congratulations! You have explored the hardest and most complicated Fastify components. It could take a while before gaining full control over these functions, so don't rush – take your time. Keep in mind that it is fine to not customize the schema controller and use the default one. This section's takeaway is how the system works under the hood, but it's still a valuable option to explore because then you will not have nasty surprises during your application development.

Now that you have learned about the Schema Controller Factory component and its configuration and customization, we can move on. Keep calm; it will be easier from now on.

Customizing the schema validator compiler

In the previous section, we explored the Schema Controller Factory, one of the two ways to substitute Fastify's schema compiler. Looking at *Figure 5.4* and the conditions to check whether Fastify must create a compiler, we can tackle the question *Does the root or the plugin instance have a compiler already?*

Customizing the schema validator is quite easy at this point:

```
app.setValidatorCompiler(myValidatorCompiler) // [1]
app.register(async function plugin (instance, opts) {
  instance.setValidatorCompiler(myValidatorCompiler) // [2]
  app.post('/schema-ref', {
    handler: echo,
    validatorCompiler: myValidatorCompiler, // [3]
    schema: {
      body: mySchema
    }
  })
})
```

The `myValidatorCompiler` variable is the same as in the previous section: it has the same interface and the same result is returned.

As you know, Fastify lets you customize the schema compiler in two different stages:

- Customize the validator compiler for the root application instance [1] or plugin instance [2]
- Customize the validator compiler for a single route through the route's options [3]

As we have just seen in the example, customizing the validator compiler forces Fastify to skip the Schema Controller Factory call. The function you provide is used instead of building it.

This type of customization is easier to apply compared to the Schema Controller Factory, and it gives you the possibility to change even the tiniest pieces of your application. The typical function of this feature is to support two different compilers in the same Fastify application. This is really useful to migrate applications written using the `joi` module. The following code shows an example:

```
app.register(async function plugin (instance, opts) {
  function joiCompiler ({ schema, method, url, httpPart })
  {
    return function (data) { return schema.validate(data) }
  }
  instance.setValidatorCompiler(joiCompiler)
  instance.post('/joi', {
    handler: echo,
    schema: {
      body: Joi.object({
        hello: Joi.string().required()
      })
    }
  })
})
```

All the routes registered in the `plugin` rely on the `joiCompiler` returned function to validate the HTTP parts. Notice that the `schema` argument is actually a `Joi.object()` instance object. It has been provided during the route registration, and it is not a standard JSON Schema at all. Fastify doesn't complain about this because you have provided a custom validator compiler, so it is okay if the provided compiler knows how to manage the input schema object.

The routes registered out of the plugin context rely on Fastify's default validator compiler! We must thank the encapsulation yet again!

Now the validation process and all its pieces have no more secrets for you. You have acquired a deep knowledge of how your routes' input is validated before executing the handler.

It is time to meet the serializer compiler, but don't worry, the concepts we explored in this extensive section will be the same in the next one.

Understanding the serialization process

Serialization is the process of transforming complex objects or primitive data into a valid data type that can be transmitted to the client. A **valid data type** is a String, a Stream, or a Buffer.

In the *Understanding validation and serialization* section, we introduced the concept of Fastify's serialization process, which uses JSON schemas to adapt a response payload to a defined format. This is the only task that this process carries out. It doesn't apply any sort of validation of the output. This is often confusing because the JSON Schema is associated with the validation phase. Therefore, it would be more appropriate to compare it to filter processing rather than to a validation.

The actors in action are quite similar to what we saw in the *Building a new validator compiler* section, with some additions. In the following diagram, *Figure 5.5*, we are going to present these additions, extending the **Serialization** box we saw in *Figure 5.3*:

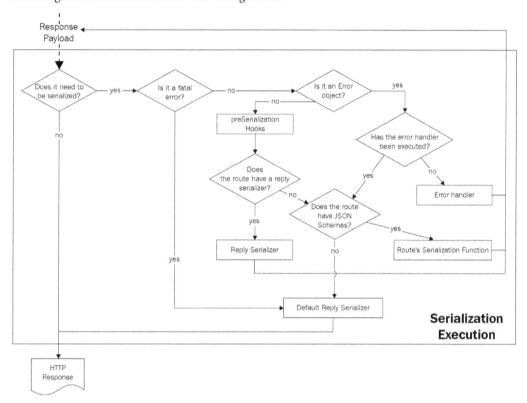

Figure 5.5 – The serialization workflow

Figure 5.5 shows the complete workflow Fastify carries out to serialize the response payload you are sending, calling `reply.send()` or executing a `return` statement in an `async` handler.

> **Too many payloads**
>
> To improve this section's clarity, we will name the endpoint's output object a **response payload**. For reference, it is the one that you give as an argument to the `reply.send()` method. The **serialized payload** is the response payload transformed to a valid data type and ready to be transmitted to the client.

The diagram might look a bit complex, but it condenses these Fastify rules:

- It is sending or returning an object that is not a String, a Buffer, or a Stream that triggers the serialization process

- Returning or throwing an error object will trigger the error handler we saw in *Chapter 3*

- Fastify applies a fallback to avoid starving requests due to a component's misuse, such as throwing an error in an error handler

Let's start analyzing *Figure 5.5* in detail. The first thing that stands out is that the serialization process is skipped when the response payload is a valid data type. This usually happens for web page rendering or for a file download.

When the response payload is not valid, it is processed by five main blocks that manage the payload:

- The `preSerialization` hooks manipulate the payload before the serialization. We saw this in action in *Chapter 4*.

- The error handler must manage the error and reply to the HTTP request. As you can see, it is executed only once, but we read about it in *Chapter 3*.

- The **Reply Serializer** is a new simple component. It must convert the object to a valid data type.

- The **Route's Serialization Function** is produced by the serializer compiler.

- The last one, the **Default Reply Serializer**, acts when there is no customization.

Thanks to the flowchart, it should be easier to navigate through the serialization process. For example, it is clear that a custom reply serializer has priority over the serialization function. But let's look at the code to learn how to use these components to serialize a response payload.

The reply serializer

This component helps you manage a response that is not JSON and needs to be serialized in a different format. It is a synchronous function that must return a valid data type (a String, or a Stream, or a Buffer). If something else is returned or thrown, a fatal error will be sent in response. The reply serializer's `this` context is the reply object itself – it may be helpful to set additional headers.

The usage is quite straightforward at this point of your path through the Fastify world:

```
function mySerializer (payload, statusCode) {
  return `<payload>${payload}</payload>`
}
app.setReplySerializer(mySerializer) // [1]
app.get('/reply-serializer', function handler (request,
reply) {
  reply.type('application/xml')
```

```
    .serializer(mySerializer) // [2]
    .send({ hello: 'world' })
})
```

You can assign a custom serializer to the root or plugin server instance [1]. As always, in this case, all the routes registered in that context will run it.

The other option is to run the custom serializer only when needed to call the `reply.serializer()` method and passing `serializer` as an argument [2]. Remember to set the content-type header in this case, or you may encounter Fastify's unpredictable results.

That is all you must know about the reply serializer. This component is used to reply with content types that are not JSON, such as XML, compression buffers, YML, and so on.

Having closed this parenthesis, we can start to complete our serialization journey by discussing the serialization function produced by the serializer compiler.

The serializer compiler

The serializer compiler builds a JavaScript function from a JSON Schema to serialize the response payload. It removes all the fields that are not declared in the schema, and it coerces the output fields' type. These are the tasks that Fastify's default serializer compiler does.

The module that compiles the JSON Schemas is `fast-json-stringify` (https://www.npmjs.com/package/fast-json-stringify).

> **The speed increment**
>
> The serialization process through a compiled function reduces the time taken to serialize the response's JSON object. The `fast-json-stringify` and Fastify modules have published a comprehensive benchmark that compares them to other frameworks too (https://github.com/fastify/benchmarks/).

But how can we use the JSON Schema to serialize the payload? Let's see an example:

```
app.post('/filter', {
  async handler (request, reply) {
    return {
      username: 'Foo',
      password: 'qwerty'
    }
  },
  schema: {
    response: {
      '2xx': {
```

```
        type: 'object',
        properties: {
          username: { type: 'string' }
        }
      }
    }
  }
})
```

The code snippet shows a new option field on the `schema` route's options object: the `response` property. It accepts a JSON object where each key must be the following:

- An HTTP status code such as `200`, `404`, or `500`

- An HTTP status code pattern looks like: `2xx`, meaning all the status codes from `200` to `299`

In this way, you can customize every response payload with a defined JSON Schema.

As you can see, the `/filter` endpoint returns a `password` field, but thanks to the response schema, it will not be sent to the client! The schema object will use the same schemas within the `$ref` keyword as the validator compiler: they share the `bucket` schemas, as said in the *How to build a new validator compiler* section.

Knowing the default serializer compiler

The `fast-json-stringify` module has its own implementation, and it doesn't support all the keywords provided by the Ajv validator. This is understandable because they have different functions. One great example you need to learn from is that, while the `maxLength` property of a string field is fundamental to validate the input, it is ignored by the default JSON serializer. Remember, the default JSON Schema serialization doesn't validate the data; it only filters and coerces the types.

Now we know how to use Fastify's default serializer compiler, but how is it possible to customize it? Let's learn about it in the next section.

Managing the serializer compiler

Just as you can customize the validator compiler, you can do the same with the serializer compiler. To complete the code snippet described in *Figure 5.5*, check out the following code:

```
const app = fastify({
  serializerOpts: { rounding: 'ceil' },
  schemaController: {
    compilersFactory: {
      buildSerializer: myFactory // [1]
```

```
      }
    }
  })
  app.addSchema({
    $id: 'http://myapp.com/string.json',
    type: 'string'
  })
  function myFactory (externalSchemas,
  serializerOptsServerOption) {
    return mySerializerCompiler // [2]
  }
  function mySerializerCompiler (routeData) {
    const { schema, method, url, httpStatus } = routeData //
    [3]
    return function serializer (responsePayload) { // [4]
      return `This is the payload ${responsePayload}`
    }
  }
```

The source code should look very familiar to you: the logic flow is the same as we discussed in the *How to build a new validator compiler* section, with minimal changes:

1. The `compilersFactory` option accepts the new `buildSerializer` property.

2. The `myFactory` function receives the `serializerOptsServerOption` input equal to the `serializerOpts` object (instead of the `ajv` one).

3. The `mySerializerCompiler` function receives `httpStatus` as route data [3]. The validator receives `httpPart`.

4. The compiler must return a synchronous function [4] that builds a string object.

The compiled function is then stored in the route's context, and it will be executed when the serialization process requires it, as we saw in *Figure 5.5*.

Finally, like the validator compiler, you can set the serializer compiler in every context you need it:

```
  app.setSerializerCompiler(mySerializerCompiler) // [1]
  app.register(async function plugin (instance, opts) {
    instance.setSerializerCompiler(mySerializerCompiler) //
    [2]
    app.post('/respose-schema', {
      handler: echo,
      serializerCompiler: mySerializerCompiler, // [3]
      schema: {
        response: {
          '2xx': mySchema,
```

```
            '5xx': myErrorSchema
        }
      }
    })
  })
```

The Fastify way to customize the compiler is consistent with the validator way:

- Customize the serializer compiler for the root application instance [1] or the plugin instance [2]
- Customize the serializer compiler for a single route through the route's options [3]

We have explored the serializer compiler without annoying you by repeating the logic behind the schema controller.

You should be able to customize Fastify's serializer and change the module that implements schema compilation by adopting the library that best fits your needs. As you have read, Fastify keeps its components consistent, adopting solid patterns that ease your learning curve and improve the framework itself.

Summary

This chapter has followed a long path inside Fastify's internals to unveil the JSON schema's power. Now you understand why defining a JSON Schema is a critical phase in your application setup. It can be a hard task, but data validation and a fast response are the two main reasons to do this.

We have looked at the JSON Schema specification's basics and how to use it in our routes, adopting the default Fastify components. We did not step back to configure these components, and now you have seen the whole process, you can control them to reach your goals.

It has not been easy. The concepts in this chapter are the most misused by developers that use Fastify. You did a great job, and I hope the diagrams have helped you follow the logic behind the validation and the serialization.

This chapter is the last theoretical one: congratulations! In many chapter's sections, we read concepts that were already discussed in the previous chapters. This is proof that the Fastify architecture is recursive in all its components.

Now, things are getting serious: in the next chapter, we will start building a real application that we'll use for the rest of the book to create a solid, reusable scaffolding project.

Part 2:
Build a Real-World Project

In this part, we will design and build a practical RESTful application from scratch, with a focus on creating highly reusable components. We will showcase the efficiency of the development process using Fastify by utilizing the appropriate components in the most proficient manner.

In this part, we cover the following chapters:

6

Project Structure and Configuration Management

Starting from this chapter, we are going to create a real-world RESTful cloud-native application from the initial project structure. No more foo/bar examples and Fastify theory. We will put into action what we have learned in the previous chapters. This will lead us to understand how to build an application.

This chapter will build a solid scaffolding structure that you may reuse for your future projects. You will be introduced to and use community packages and create your own plugins when needed.

This is the learning path we will cover in this chapter:

- Designing the application structure
- Improving the application structure
- Debugging your application
- Sharing the application configuration across plugins
- Using Fastify plugins

Technical requirements

As mentioned in the earlier chapters, you will need the following:

- A working Node.js 18 installation.
- The VS Code IDE from `https://code.visualstudio.com/`.
- A working Docker installation from `https://docs.docker.com/get-docker/`.
- A Git repository is highly recommended but not mandatory. Get it from `https://git-scm.com/`.
- A working command shell.

All the snippets in this chapter are on GitHub at `https://github.com/PacktPublishing/Accelerating-Server-Side-Development-with-Fastify/tree/main/Chapter%206`.

Designing the application structure

In this chapter, we will design a backend application that will expose some RESTful APIs.

The application structure enables you to write applications that are easy to implement, evolve, and maintain. A good system must be flexible to your needs and the application's changes. Additionally, it should impose some implementation design to let you and your team avoid some major pitfalls, which could lead to an unstable and untestable application.

This chapter will discuss the application's features only marginally. In fact, a scaffolding project should not care about them, but it should apply them to any project. For this reason, we will create an application with some health check routes and a MongoDB connection ready to be used.

We will introduce a set of Fastify plugins that will help us structure our application and reduce the burden of writing some utilities from scratch, which have already been developed and tested on production by multiple projects.

The critical takeaway here is to understand *why* we will build the following structure in this manner. The structure we are going to see is not mandatory, and you might be critical of the proposed design. We think it is important to personalize the application to adapt it to your own needs and preferences.

We are done with the talking. Let's start to build our application!

Setting up the repository

Before we start writing the code, we should define a baseline from where to start. Thankfully, building an empty Fastify application is quite easy due to an official utility built by the Fastify team. To use it, we need to open the command shell and type the following commands:

```
mkdir fastify-app
cd ./fastify-app
npm init fastify
```

Running these commands creates a new `fastify-app` folder and executes the `npm init` command from within it.

> **The init command**
>
> When you run the `init` command, npm executes the `create-fastify` module, which you can find at this GitHub repository: `https://github.com/fastify/create-fastify`. You can create the `create-my-app` authority to build an application scaffolding to speed up the project initialization.

As a result, you will see the following files and directories:

- `package.json`: This is the project entry point.
- `app.js`: This is the main application file. It is the first file to be loaded.
- `plugins/`: This folder stores custom plugins. It contains some sample files.
- `routes/`: This folder stores the application's endpoints. It includes a few example endpoints.
- `test/`: This is the folder where we write our application's test.

The starter application is ready to be installed by running `npm install`. After the installation, you may find these scripts that are already configured useful:

- `npm test`: This script runs the test of the scaffolding application.
- `npm start`: This script will start your application.
- `npm run dev`: This script will start your application in *development mode*. The server automatically reloads your application at every file change.

We have just built a basic Fastify application setup that we will customize to create our application. Reading the generated source code using the `init` command, you will find comments that help you orientate yourself, giving you some insight that we will see throughout this chapter.

Application versioning

To start building a real-world application, it is essential to set up a **version control system (VCS)**. This software lets us version our source code and manage changes. You should use **Git** software for this task as it is the standard de facto software in the tech industry. However, learning about Git is not the aim of this book. To find out how to install and use Git, check out the *Technical requirements* section.

At this stage, all the commands we mentioned in this section should work on your PC. Take some time to try the development mode by editing the `routes/root.js` file, and adding a new `GET / hello` route while the server is up and running!

Understanding the application structure

The application structure we have built so far has excellent out-of-the-box features. It is founded on some of the following pillars:

- It relies on the `fastify-cli` plugin to start the application and provide the developer mode
- It takes advantage of the `@fastify/autoload` plugin to automatically load all the files contained in the `plugins/` and `routes/` folders

It is not mandatory to use these two plugins, but they offer many great features that we will need at some point during our application evolution. Using them now will help you gain confidence in those features and speed you up later.

The fastify-cli plugin

The `fastify-cli` **command line interface (CLI)** helps us start our application. It is used on the `package.json` file. The `scripts` property uses the `fastify start` command with some options to create the `app.js` file. As you will notice, the `app.js` file exports a typical Fastify `async function (fastify, opts)` plugin interface. The file is loaded by the CLI as a usual `app.register()` call, as we saw in the *Adding a basic plugin instance* section of *Chapter 1*. In this case, we are not instantiating the Fastify root server instance or calling the `listen` method. All these tasks are accomplished by `fastify-cli`, saving us from the code boilerplate.

Moreover, the CLI improves the development process by implementing proper settings and options:

- It adds *a graceful shutdown*, as we read about in the *Shutting down the application* section in *Chapter 1*.
- It reads a root `.env` file by default. It is a `key=value` file that only contains string. It is used to describe the settings that will be read from the operating system environment setup. All these variables are mapped into the Node.js `process.env` object.
- It starts listening on the `PORT` environment variable for all the hosts.
- It accepts the `--debug` option to start the application in *debug mode* to debug your code.
- It exposes the `--options` flag to customize the Fastify server option since we are not instantiating it directly.
- The `--watch` argument turns on server auto-reloading when a file in the project changes.
- The `--pretty-logs` argument makes the output logs readable on the shell.

You can find the detailed documentation on the CLI in the repository here: `https://github.com/fastify/fastify-cli#options`.

We are going to customize our `fastify-cli` installation in the next section: *Improving the application structure*.

The @fastify/autoload plugin

The autoload plugin automatically loads the plugins found in a directory and configures the routes to match the folders' structure. In other words, if you create a new `routes/test/foo.js` file with the following content, a new `GET /test/` route will be declared:

```
module.exports = async function (fastify, opts) {
  fastify.get('/', async function (request, reply) {
```

```
    return 'this is an example'
  })
}
```

This behavior follows a **convention over configuration** design paradigm. Its focus is reducing the implementation to provide the desired behavior by default.

By using the `@fastify/autoload` plugin, you can see every file as an encapsulated context, where every folder composes the plugin's `prefix`, as we saw in *Chapter 2*.

Given that we don't need the entire code to be autoloaded, it is fine to require our plugins and routes to register the required parts as children, but we need to find a clear pattern for doing this.

The autoload plugin has some default behaviors and is based on a naming convention that could be confusing if you don't study the documentation. We will not explore all the different options and combinations: there are a lot, and it would take a book to explain them! For this reason, we will customize the default application to make it more transparent and easy to maintain, using the most consolidated setup.

Improving the application structure

We are becoming confident with our basic application setup, but it is not yet ready. We can improve it to define our own rules and give a better developer experience. We will create a solid project scaffolding focusing on the big picture first. After that, we will complete the Fastify setup details.

Starting an optimal project

A good project is not only fancy technology in action but it must also provide a good developer experience, reducing any burden. According to Fastify's philosophy, we should set up our new code base keeping this aspect in mind. Therefore, we will provide a brief introduction to those aspects, as they are priceless and time-saving but often underrated.

The README file

The first addition to our project is a `README.md` file. A typical readme file introduces newcomers to an existing project, answering a set of basic information questions as follows:

- What are the project's requirements? Do you need a database or some other external resources?
- How do you install the application? What package manager and Node.js version does it use?
- How do you start the application? Where can missing data (such as environment variables) be found?
- How do you develop the application? Are there some conventions that developers must follow?
- How do you test the application? Does it require unit tests or end-to-end tests?

- How do you deploy the application? What are the processes and the environment's URLs?

- What should a developer do if there is missing information?

This set of questions may not have answers at the beginning of the project, but it is good to note the unanswered ones and to find answers to them in the future.

> **Positive vibes**
>
> The README file has many other positive effects on the team's morale. A developer will feel productive after reading it. We suggest you read the following article if you want to know more about the importance of the README file. The article is written by Tom Preston-Werner, co-founder of GitHub and many other open source projects: `https://tom.preston-werner.com/2010/08/23/readme-driven-development.html`.

The most important thing is to keep the README file up to date. Every reader should improve it by adding missing parts or removing old ones. If this descriptive file becomes too large, it will become practically unreadable. So consider creating a `docs/` folder to split it into more readable pieces.

The code linter

Another essential step you should consider is adopting a **linter**. A linter is a piece of software that analyzes the source code statically and warns you about possible issues or typos to save you hours of debugging and adding `console.log` statements.

If you don't want to choose a linter and configure it, we suggest opting for `standard`. It is a zero-configuration linter, which can be installed by running `npm install standard --save-dev`, and is ready to be integrated into custom `package.json` scripts, as follows:

```
"scripts": {
  "lint": "standard",
  "lint:fix": "standard --fix",
// file continues
```

In this way, we will be able to run `npm run lint` to check our source code and get feedback. If there is no output, it means that all is good! Running `npm run lint:fix` will fix the errors automatically when possible – it is helpful for formatting issues.

Note that we can integrate the lint validation with a requirement for the project. We need to modify the `package.json` scripts like so:

```
"pretest": "npm run lint",
"test": "tap \"test/**/*.test.js\"",
```

The npm test command will automatically execute the pretest and posttest scripts if present. You may find it helpful to read the npm pre and post documentation to run extra commands before and after a script execution: https://docs.npmjs.com/cli/v7/using-npm/scripts#pre--post-scripts.

The container build

The Fastify application we will build during this book can work on a **production server** or a **container engine**. The former can run our application using the canonical npm start command. The latter requires a **container** to run our application.

We are not going to go into much depth about the container topic since it is out of the scope of this book. It is useful to enter this topic using a secure example, which you will see next. This configuration is ready to build Node.js production containers.

To build a Docker container that contains our software, we must create a Dockerfile document on the root directory. This file contains the instruction to build our container image in the most secure way possible:

```
FROM node:18-alpine as builder
WORKDIR /build
COPY package.json ./
COPY package-lock.json ./
ARG NPM_TOKEN
ENV NPM_TOKEN $NPM_TOKEN
RUN npm ci --only=production --ignore-scripts
FROM node:18-alpine
RUN apk update && apk add --no-cache dumb-init
ENV HOME=/home/app
ENV APP_HOME=$HOME/node/
ENV NODE_ENV=production
WORKDIR $APP_HOME
COPY --chown=node:node . $APP_HOME
COPY --chown=node:node --from=builder /build $APP_HOME
USER node
EXPOSE 3000
ENTRYPOINT ["dumb-init"]
CMD ["npm", "start"]
```

The previous script snippet defines how Docker should create the container. These steps are described as follows:

1. FROM: This starts the multistage build of our application from a base image within Node.js and with npm installed.

2. ENV: This defines some useful environment variables that will always be set up in the container.

3. `COPY`: This copies the `package.json` files into a container's folder.

4. `WORKDIR`: This sets the current working directory and where to run the subsequent commands from.

5. `RUN npm ci`: This installs the project dependencies using the `package-lock` file.

6. `COPY`: This copies the application source code into the container.

7. `RUN apk`: This installs the *dumb-init* software into the container.

8. `USER`: This sets the container's default user at runtime. This user is a least-privilege user to secure our production environment.

9. `EXPOSE`, `ENTRYPOINT`, and `CMD`: These define the container's external interface and set the application start as the default command on container initialization.

This file is a secure and complete descriptor to build the application container, and it is the perfect baseline for our project. It will change over time as the logic to start up the application changes.

We adopted a multistage build because you may need to provide some secrets to your application to install it successfully. A typical example is to rely on a private npm registry. These secrets must not be persisted into the application's Docker image, else anyone who gets access to the Docker image will be able to leak the npm token and get access to your private npm registry. The multistage build, instead, consists of a two step build process:

1. Create a `builder` image that has access to the private npm registry, and download the application's dependencies.

2. Copy the dependencies from the `builder` image to the application one, and then throw away the `builder` image and its secrets.

Finally, to use this file you must run the `docker build -t my-app` command. and the build process will start. We will discuss this topic further in *Chapter 10*.

The test folder

We must not forget about testing our application. As a reminder, we should create a `test/` folder containing all the application's tests, which we will implement in *Chapter 9*. However, we need to be comfortable with our application structure first. This is because the test implementation depends on the project implementation. Only after we have reached a stable solution will we be able to write our basic assertions, such as the following:

*   The application starts correctly

*   The configurations are loaded in the right order

The tests' assertions must reply to these questions positively to prove that our configuration scaffolding works as expected.

We have completed the basic project setup. It was not strictly about Fastify, but it concerned how to start a good project even before writing some code lines. We must not forget that every line of the code or documentation will become the future legacy code. For this reason, we should do our best to make it easy and secure to use.

Now let's see how to continue the code base structure to prepare us to write our first real-world application.

Managing project directories

The application scaffolding currently has two main directories loaded in the same manner by the @ fastify/autoload plugin. But these folders are not equal at all.

We need more buckets to order the source code and to make it clear and readable. For example, we saw that the JSON schemas structures can become verbose, so we should move them from our routes implementation to keep us focused on the business logic.

We are going to create new project folders to define the final project structure and declare what they should contain. Every directory will have its own @fastify/autoload configuration.

The folders are presented in order of loading, and you must not forget that while writing:

```
fastify.register(AutoLoad, { dir: path.join(__dirname, 'plugins') })
fastify.register(AutoLoad, { dir: path.join(__dirname, 'routes')  })
```

The code will load the plugins first and then the routes. If the process was reversed, it would lead to errors. For this reason, the following project folders are presented in order of loading.

Loading the plugins

The plugins/ folder should contain plugins that are based on the fastify-plugin module. We encountered this module in *Chapter 2*.

You should store all the application's components in this directory, which need to be registered as the following:

- As a root server instance's plugin, such as a database connection.
- As reusable plugin components. These files are not intended to be autoloaded, but they must be registered when required by routes. An example could be an authentication plugin.

For this reason, a good approach is to edit the plugins folder's autoload setup as follows:

```
fastify.register(AutoLoad, {
  dir: path.join(__dirname, 'plugins'),
  ignorePattern: /.*.no-load\.js/,
  indexPattern: /^no$/i,
  options: Object.assign({}, opts)
})
```

With this new setup, the `ignorePattern` property lets us ignore all those files that end with the `.no-load.js` filename. This option tells us at first sight what is being loaded and what is not, improving the project's clarity. Note that the pattern does not consider the directory name.

> **Customize as per your preferences**
>
> If you don't like the "no-load" pattern, you can invert the logic by setting the `ignorePattern:` `/^((?!load\.js).)*$/` property value, and load only those files with the `.load.js` suffix.

The `indexPattern` property instead disables a `@fastify/autoload` plugin. By default, if a directory contains an `index.js` file, it will be *the only one loaded*, skipping all other files. This could be an undesired behavior that the `indexPattern` option prevents.

Finally, the `options` property lets us provide a configuration object as input for the plugins that are being loaded. Let's take the `plugins/support.js` file as an example. It exports the `module.exports = fp(async function (fastify, opts)` interface. The autoload's `options` parameter matches with the `opts` argument. In this way, it is possible to provide a configuration for all the plugins. We will go deeper into this aspect in the *Loading the configuration* section.

Loading the schemas

The JSON schemas are a crucial part of a secure project and need a proper stage on the application's structure. Creating a `schemas/` folder to store all the JSON schemas is convenient while the application develops: you will find out soon that we will work with numerous schemas.

In the folder, we will add a `loader.js` file that has one task. It must add all the JSON schemas we will need for our application:

```
const fp = require('fastify-plugin')
module.exports = fp(function (fastify, opts, next) {
  fastify.addSchema(require('./user-input-headers.json'))
  next()
})
```

The code snippet is actually a plugin, but it may become bigger and bigger. Isolating it from the `plugins/` folder lets us avoid the chaotic, infinite scrolling when navigating the code base.

There will be many schemas to work with because it is *highly recommended to define a schema for each HTTP part* you need to validate or serialize. All the HTTP verbs require different validation types; for example, an `id` field should not be submitted as input on a POST route, but it is mandatory in a PUT route to update the associated resource. Trying to fit a general JSON schema object into multiple HTTP parts may lead to unexpected validation errors.

There is automation to load the schemas into a directory. So we will need to list all the files in the current directory and run the `addSchema` method. We will see how to implement it in *Chapter 7*.

To load the `loader.js` file into our project, the `@fastify/autoload` plugin may seem like killing a fly with a sledgehammer, but it is a good method to use to split our schemas even more. Register the plugin in the `app.js` file, as follows:

```
fastify.register(AutoLoad, {
  dir: path.join(__dirname, 'schemas'),
  indexPattern: /^loader.js$/i
})
```

In this way, the autoload plugin will exclusively load the `loader.js` files created in the directory tree. So we will be able to make the subfolders, such as the following:

- `schemas/headers/loaders.js`
- `schemas/params/loader.js`
- `schemas/body/loader.js`

Similarly, we can make more subfolders for every HTTP method, and you will find the best tree structure that fits your application. We found that splitting the schemas by HTTP parts is the best way of ordering. This division speeds up our source navigation even more. We will be able to create some utilities dedicated to each HTTP part, such as some regular expressions for the headers and complex reusable objects for the body input.

Loading the routes

The `routes/` folder contains the application's endpoints. Still, all the files are loaded automatically, making it hard to split the code base into smaller pieces. At this point, a `utility.js` file will be loaded by `@fastify/autoload`. Moreover, defining an `index.js` file will prevent the loading of other files, as we saw previously in the *Loading the plugins* section.

The best rules we suggest applying when it comes to the `routes/` folder are the following:

- Autoload only those files that end with `*routes.js`. Discard all the other files in the folder. The ignored files can be registered manually or used as utility code.
- We must not use the `fastify-plugin` module in this folder. If we need it, we should stop and think about whether that code could be moved to the `plugins/` folder.
- Turn on the **autohook** feature in `@fastify/autoload`.

To apply these rules, we need to set up the autoload plugin, as shown in the next code example:

```
fastify.register(AutoLoad, {
  dir: path.join(__dirname, 'routes'),
  indexPattern: /.*routes(\.js|\.cjs)$/i,
  ignorePattern: /.*\.js/,
  autoHooksPattern: /.*hooks(\.js|\.cjs)$/i,
```

```
    autoHooks: true,
    cascadeHooks: true,
    options: Object.assign({}, opts)
  })
```

The code snippet introduces two new parameters. The `autoHooks` flag lets you register some hooks for every `routes.js` file. The `cascadeHooks` option also turns this feature on for the subdirectories.

Let's make an example within this structure, which we will further discuss in *Chapter 8*. Given this folder tree, the `authHooks.js` file exports the standard Fastify plugin interface, but it must only configure life cycle hooks:

```
routes/
└┬ users/
 ├── readRoutes.js
 ├── writeRoutes.js
 ├── authHooks.js
 └┬ games/
  └── routes.js
```

This example configures some `onRequest` hooks to check the client's authorization. Now, would you expect `routes/games/routes.js` to be authenticated?

If your answer is yes, `cascadeHooks: true` is right for you. We think most of you naturally find that the hooks registered as `autoHooks` in the parent's folder are added to the children.

If the answer is no, you can change the `cascadeHooks` option to `false` to avoid adding the hooks to the children folders. In this case, the hooks are only loaded for the `readRoutes.js` and `writeRoutes.js` files. You may need to duplicate the `authHooks.js` file for every folder you need the authentication for or register it as `plugins/` in the root context.

All these rules should be listed in the `README.md` file to shout out how to develop a new set of routes. This will help your team to join the project without studying all of Fastify's plugins in detail.

Now, we have a clear structure ready to welcome your business logic endpoints. Still, before moving forward, we need to face the last aspect of the repository structure: how to manage the configuration.

Loading the configuration

The configuration is the first step that our application must execute to start correctly. In the *Understanding configuration types* section of *Chapter 1*, we discussed three types of configuration that our application needs:

- **Server options**: This is the Fastify root instance setup

- **Application configuration**: This is the extra setting that sets how your application works

- **Plugin configuration**: This provides all the parameters to configure the plugins

These configurations have different sources: a server option is an object, the plugins' configurations are complex objects, and the application configuration relies on the environment. For this reason, they need to be treated differently.

Loading the server configuration

The server instantiation is not under our control. The `fastify-cli` plugin does it for us, so we need to customize it to set the **server options**.

We must edit the `start` and `dev` scripts into our `package.json`:

```
"start": "fastify start -l info --options app.js",
"dev": "npm run start -- --watch --pretty-logs"
```

We have modified the `dev` script to execute the `start` one in order to reduce code duplication and avoid copy-and-paste errors. The double dash (`--`) lets us forward extra arguments to the previous command. So, it is like appending parameters to the `start` script.

Adding the `--option` flag in the `start` script equals adding it to both commands without replicating it.

The `--option` flag uses the `app.js` options property. Let's not forget to add the server options we would like to provide during the Fastify initialization and place them at the bottom of the file:

```
module.exports.options = {
  ajv: {
    customOptions: {
      removeAdditional: 'all'
    }
  }
}
```

In doing so, we are exporting the same JSON object we were providing to the Fastify factory. Restarting the server will load these settings. A sharp developer may notice that we did not configure the `logger` options, but we can see logs in our console. This happens because our customization is merged within the `fastify-cli` arguments, and the `-l info` option sets the log level to `info`.

Centralizing all the configurations into one place is best practice, so remove the `-l` argument from the `package.json` script and add the usual `logger` configuration into the exported JSON.

For the sake of centralization, we can move `app.js module.exports.options` into a new dedicated `configs/server-options.js` folder. The server option does not need any async loading, and it can read the `process.env` object to access all the `.env` file values loaded at startup by `fastify-cli`.

Loading the application configuration

The application configuration is mandatory for every project. It tracks secrets, such as API keys, passwords, and connection URLs. In most cases, it is mixed across sources, such as filesystems, environment variables, or external **Secret Managers**, which store that information securely, providing additional control over the variables' visibility.

We will focus on one basic loading type: the environment variable. This is the most common type, and it enables us to use Secret Managers. To go deeper into understanding external Secret Managers, we suggest you read this exhaustive article: `https://www.nearform.com/blog/uncovering-secrets-in-fastify/`. It explains how to load configuration from the most famous providers, such as AWS, Google Cloud Platform, and Hashicorp Vault.

The environment configuration is tied to the system where the software is running. In our case, it could be our PC, a remote server, or a colleague's PC. As previously mentioned, Node.js loads all the **operating system** (**OS**) environment variables by default into the `process.env` object. So, working on multiple projects could be inconvenient as it would change the OS configuration every time. Creating Replace with an `.env` text file in the project's root folder is the solution to this annoying issue:

```
NODE_ENV=development
PORT=3000
MONGO_URL=mongodb://localhost:27017/test
```

This file will be read at the startup by the fastify-cli plugin, and it will be ready to access. Note that this file will overwrite the `process.env` property if it already exists. So you have to make sure that what is in your `.env` file is the application configuration's source of truth. You can verify the correct loading of the `app.js` file by adding a simple log:

```
module.exports = async function (fastify, opts) {
    fastify.log.info('The .env file has been read %s',
    process.env.MONGO_URL)
```

Since the `.env` file may contain sensitive data, you should never commit it to your repository. In its place, you should commit and share a `.env.sample` file. It lists all the keys that must be set as environment variables without any secret values.

Saving sensitive data in the repository is dangerous because whoever has access to it may access upper environments such as production. In this way, the security is moved from access to an environment, such as a server, to the Git repository setting. Moreover, if an environment variable needs to be updated, you should commit it and publish new software version to deploy the change in other environments. This is not correct: the software must not be tied to the environment variable values. Remember to track all files that must be secrets to the `.gitignore` and `.dockerignore` files.

We can improve the `.env.sample` file to make it as straightforward as possible. For this, we need the `@fastify/env` plugin, which throws an error whenever it doesn't find an expected variable.

First of all, we need a JSON schema that describes our .env file. So, we can create the schemas/dotenv.json file:

```json
{
  "type": "object",
  "$id": "schema:dotenv",
  "required": ["MONGO_URL"],
  "properties": {
    "NODE_ENV": {
      "type": "string",
      "default": "development"
    },
    "PORT": {
      "type": "integer",
      "default": 3000
    },
    "MONGO_URL": {
      "type": "string"
    }
  }
}
```

The JSON schema environment is quite linear. It defines a property for each variable we expect. We can set a default value and coerce the type as we did for the PORT property in the JSON schema. Another nice takeaway is the $id format. It has a **Uniform Resource Name** (**URN**) syntax. The specification discussed in *Chapter 5* explains how it can be a **Uniform Resource Identifier** (**URI**). A URI may be a URL that identifies a location or a URN when it identifies a resource name without specifying where to get it.

Now, we must not forget to update the schemas/loader.js file by writing fastify.addSchema(require('./dotenv.json')) to load the schema.

To integrate the @fastify/env plugin, we are going to create our application's first plugin, plugins/config.js:

```js
const fp = require('fastify-plugin')
const fastifyEnv = require('@fastify/env')
module.exports = fp(function (fastify, opts, next) {
  fastify.register(fastifyEnv, {
    confKey: 'secrets',
    schema: fastify.getSchema('schema:dotenv')
  })
  next()
}, { name: 'application-config' })
```

The plugin will be loaded by the autoload feature, so we need to run the server and try it out. Starting the server without the MONGO_URL property will stop the startup and inform you that the key is missing. Moreover, it will add a decorator to the Fastify instance, named the confKey value. So the .env keys will be available to read the fastify.secrets property, decoupling the code from the global process.env object.

Before going further into the plugins' configuration loading, we should pay attention to the name option we just set as input to the fp function. It will be the key to understanding the *Loading the plugins' configurations* section.

Loading the plugins' configurations

The plugin's settings are dependent on the application's configuration. It requires the application's secrets to configure itself to work as expected, but how can we configure it?

Before proceeding, you will need a MongoDB instance up and running in your development environment. We will use this database in future chapters. You can download the community edition for free at https://www.mongodb.com/try/download/community or use a temporary Docker container starting it with the following command:

```
docker run -d -p 27017:27017 --rm --name fastify-mongo mongo:5
docker container stop fastify-mongo
```

These Docker commands start and stop a container for development purposes. The data you store will be lost after the shutdown, making it suitable for our learning process.

> **Track all commands**
>
> It is best to store all the useful commands in the package.json scripts property to run the project correctly. In this way, whether you choose a Docker container or a local MongoDB installation, you will be able to run npm run mongo:start to get a running instance ready to use.

After the MongoDB setup, let's integrate the @fastify/mongodb plugin. It provides access to a MongoDB database to use on the application's endpoints. You need to install it by running the npm install @fastify/mongodb command, then create a new plugins/mongo-data-source.js file:

```
const fp = require('fastify-plugin')
const fastifyMongo = require('@fastify/mongodb')
module.exports = fp(async function (fastify, opts) {
  fastify.register(fastifyMongo, {
    forceClose: true,
    url: fastify.secrets.MONGO_URL
  })
```

```
}, {
  dependencies: ['application-config']
})
```

In the code snippet, the MongoDB configuration is contained in the plugin's file itself, but it requires the application's configuration to load correctly. This is enforced by the `dependencies` option. It is an `fp` function argument that lists all the plugins that have previously been loaded. As you can see, the `name` parameter we set in the previous *Loading the application configuration* section, gives us control over the plugins' loading order.

We are creating a solid and clear project structure that will help us in our daily job:

- It suggests to us where the code should be written

- It enforces the use of the plugin system, improving our source code to rely on encapsulation instead of global variables and side effects

- It enforces declaring the dependencies between plugins to avoid mistakes and control the load order

Can we take advantage of the Fastify architecture within this structure? If things go wrong, how can we handle mistakes? Let's have a look at this next.

Debugging your application

In a real-world application, errors may happen! So, how can we handle them within our structure? Yes, you guessed right, with a plugin!

In production, the logs files are our debugger. So, first of all, it is important to write good and secure logs. Let's create a new `plugins/error-handler.js` plugin file:

```
const fp = require('fastify-plugin')
module.exports = fp(function (fastify, opts, next) {
  fastify.setErrorHandler((err, req, reply) => {
    if (reply.statusCode >= 500) {
      req.log.error({ req, res: reply, err: err },
      err?.message)
      reply.send(`Fatal error. Contact the support team. Id
      ${req.id}`)
      return
    }
    req.log.info({ req, res: reply, err: err },
    err?.message)
    reply.send(err)
  })
  next()
})
```

The code snippet is a customization of the error handler we learned about in *Chapter 3*. Its priorities are to do the following:

- Log the error through the application's logger

- Hide sensible information when an unexpected error happens and provide helpful information to the caller and contact the support team effectively

These few lines of code accomplish a great deal, and they can be customized even more to adapt to your needs, such as adding message error internationalization.

After the error handling baseline we just set, we can use an IDE debugger to spot and solve issues. To do so, we need to edit the application's `package.json` script:

```
"dev": "npm run start -- --watch –pretty-logs -debug"
```

This will start the Node.js process in debug mode. At the startup, we will read a message in the console that informs us how to attach a debugger:

```
Debugger listening on ws://127.0.0.1:9320/da49367c-fee9-42ba-b5a2-
5ce55f0b6cd8
For help, see: https://nodejs.org/en/docs/inspector
```

Using VS Code as an IDE, we can create a `.vscode/launch.json` file, as follows:

```
{
  "version": "0.2.0",
  "configurations": [
    {
      "name": "Attach to Fastify",
      "type": "node",
      "request": "attach",
      "port": 9320
    }
  ]
}
```

Pressing *F5* will connect our debugger to the Node.js process, and we will be ready to set breakpoints and check what is happening in our application to fix it.

Now, you have seen how fast and smooth configuring Fastify is to accomplish demanding tasks!

You have improved your control over the code base, and we will now see a new technique to manage the application's configuration. You will master the project scaffolding setup as these skills build up.

Sharing the application configuration across plugins

The *Loading the plugins' configurations* section discussed how a plugin could access the application configuration. In this case, the plugin accesses the `fastify.secret` plain object to access the environment variables.

The configuration may evolve and become more complex. But, if you just intended to centralize the whole plugin's settings into a dedicated plugin, how could you do that?

We can modify the `config.js` plugin and move it to the `configs/` directory. By doing this, we are not loading it automatically anymore. Then, we can integrate the `@fastify/mongodb` configuration:

```
module.exports = fp(async function configLoader (fastify, opts) {
  await fastify.register(fastifyEnv, {
    confKey: 'secrets',
    schema: fastify.getSchema('schema:dotenv')
  })

  fastify.decorate('config', {
    mongo: {
      forceClose: true,
      url: fastify.secrets.MONGO_URL
    }
  })
})
```

In the code snippet, you can see the following main changes:

- The plugin exposes the `async` interface.
- The `@fastify/env` plugin's register is awaited to execute the plugin. In this way, `fastify.secrets` will be immediately accessible.
- A new decorator has been added to the Fastify instance.
- The plugin no longer has the `name` parameter. Since we are going to load it manually, a name is not necessary. In any case, it is a good practice to leave it: we want to show you that we are breaking the bridges between the `mongo-data-source.js` and `config.js` files.

These changes are breaking our setup due to the following reasons:

- The `config.js` file is not loaded
- The `mongo-data-source.js` file relies on `fastify.secrets`

To fix them, we need to edit app.js, as follows:

```
await fastify.register(require('./configs/config'))
fastify.log.info('Config loaded %o', fastify.config)
```

These lines must be added after the autoload schemas configuration because we validate the environment variable through the schema:dotenv schema.

After that, we can update the plugins' autoload options as follows:

```
fastify.register(AutoLoad, {
  dir: path.join(__dirname, 'plugins'),
  dirNameRoutePrefix: false,
  ignorePattern: /.*.no-load\.js/,
  indexPattern: /^no$/I,
  options: fastify.config
})
```

Finally, we can fix the mongo-data-source.js file by removing a lot of code:

```
module.exports = fp(async function (fastify, opts) {
  fastify.register(fastifyMongo, opts.mongo)
})
```

As you can see, it has become much lighter. We have removed the dependencies parameter as well because we don't want to access the fastify.secret decorator.

This change has a significant impact on the code logic. With this code restyle, the mongo-data-source.js file is decoupled from the rest of the application because all the settings are provided by the input opts argument. This object is provided by the @fastify/autoload plugin, mapping the options parameter.

You now have a comprehensive and solid knowledge of the configuration and how to best manage it. You can use the previous code example to become confident in tweaking the plugins and playing within the autoload plugin. You will find that the source code in the book's repository adopts the first solution we saw in the *Loading the plugins' configurations* section.

To complete the project scaffolding, we need to add a few more features that are key pieces to consider this basic structure solid and ready to use for our development process. We will learn about some new plugins that add these missing capabilities to our application.

Using Fastify's plugins

The project structure is almost complete. Fastify's ecosystem helps us improve our scaffolding code base with a set of plugins you will want to know about. Let's learn about them and add them to the plugins/ folder.

How to get a project overview

Documenting a complete list of all the application's endpoints is a tedious task, but someone in the team still has to do it. Luckily, Fastify has a solution for this: the `@fastify/swagger` and `@fastify/swagger-ui` plugins.

You can integrate it by creating a new `plugins/swagger.js` file:

```
module.exports = fp(async function swaggerPlugin (fastify, opts) {
  fastify.register(require('@fastify/swagger'), {
    swagger: {
      info: {
        title: 'Fastify app',
        description: 'Fastify Book examples',
        version: require('../package.json').version
      }
    }
  })
  fastify.register(require('@fastify/swagger-ui'), {
    routePrefix: '/docs',
    exposeRoute: fastify.secrets.NODE_ENV !== 'production'
  })
}, { dependencies: ['application-config'] })
```

The previous code will register the plugins. It will automatically create the `http://localhost:3000/docs` web pages that will list all the application's endpoints. Note that the documentation will be published only if the environment is not in production, for security reasons. The API interfaces should be shared only with those people that consume them.

Note that **Swagger (OAS 2.0)** and the former **OpenAPI Specification (OAS 3.0)** define a standard to generate API documentation. You may find it interesting to learn more about it by visiting `https://swagger.io/specification/`.

How to be reachable

One of the most common issues in implementing backend APIs is the **cross-origin resource sharing (CORS)** settings. You will hit this problem when a frontend tries to call your endpoints from a browser, and the request is rejected.

To solve this issue, you can install the `@fastify/cors` plugin:

```
const fp = require('fastify-plugin')
module.exports = fp(async function (fastify, opts) {
  fastify.register(require('fastify-cors'), {
    origin: true
  })
})
```

Note that the example code is configured to let your APIs be reachable by any client. Explore this aspect further to set this plugin correctly for your future applications.

Summary

In this chapter, you have created a complete Fastify project scaffolding that will be the base structure for your following projects. You can now start a new Fastify application from scratch. You can also control all the configurations your code base needs, regardless of where they are stored.

We have looked at some of the most used and useful Fastify plugins to enhance the project structure and ergonomics. You now know how to use and customize them and that there are infinite combinations.

The solid and clean structure we have built so far will evolve throughout the course of the book. Before investing more time in the structure, we need to understand the application business logic. So, get ready for the next chapter, where we will discuss how to build a RESTful API.

7
Building a RESTful API

In this chapter, we will build upon the scaffolding structure we created in the previous chapter and dive into writing the essential parts of our application.

We will start by defining the routes of our application and then move on to connecting to data sources. We will also implement the necessary business logic and learn how to solve complex everyday tasks that we may encounter while developing a real-world Fastify application.

The chapter will be divided into several main headings, starting with defining the routes, then connecting to data sources, implementing the routes, securing the endpoints, and applying the **Don't Repeat Yourself** (**DRY**) principle to make our code more efficient.

By the end of this chapter, we will have learned the following:

- How to declare and implement routes using Fastify plugins
- How to add JSON schemas to secure the endpoints
- How to load route schemas
- How to use decorators to implement the DRY pattern

Technical requirements

To follow along with this chapter, you will need these exact technical requirements, mentioned in the previous chapters:

- A working Node.js 18 installation (`https://nodejs.org/`)
- A VS Code IDE (`https://code.visualstudio.com/`)
- An active Docker installation (`https://docs.docker.com/get-docker/`)
- A Git repository – recommended but not mandatory (`https://git-scm.com/`)
- A working command shell

All the code snippets for this chapter are available on GitHub at `https://github.com/PacktPublishing/Accelerating-Server-Side-Development-with-Fastify/tree/main/Chapter%207`

So, let's get started and build a robust and efficient application that we can use as a reference for future projects!

Application outline

In this section, we will start building a RESTful to-do application. The application will allow users to perform **Create**, **Read**, **Update**, and **Delete** (**CRUD**) operations on their to-do list, using HTTP methods such as `GET`, `POST`, `PUT`, and `DELETE`. Besides those operations, we will implement one **custom action** to mark tasks as "done."

> **What is RESTful?**
>
> **Representational State Transfer** (**RESTful**) is an architectural style to build web services that follow well-defined constraints and principles. It is an approach for creating scalable and flexible web APIs that different clients can consume. In RESTful architecture, resources are identified by **Uniform Resource Identifiers** (**URIs**). The operations performed on those resources are based on predefined HTTP methods (`GET`, `POST`, `PUT`, `DELETE`, etc.). Every call to the API is stateless and contains all the information needed to perform the operation.

Fastify is an excellent choice to develop RESTful APIs, due to its speed, flexibility, scalability, and developer-friendliness. In addition, as we saw in previous chapters, its modular plugin architecture makes it easy to add or remove functionality as needed, and its low-level optimizations make it a robust choice for high-traffic applications. Finally, we will take advantage of this architecture to organize our code base in a scoped way, making every piece of it independent from the others.

Defining routes

Let's start evolving the application from our basic project structure by adding a new plugin that defines our RESTful routes. However, we will not implement the logic of single routes right now, since we first need to look at the data source in the forthcoming *Data source and model* section.

The following `routes/todos/routes.js` snippet defines the basic structure of our routes plugin:

```
'use strict'

module.exports = async function todoRoutes (fastify, _opts) { // [1]
  fastify.route({
    method: 'GET',
    url: '/',
    handler: async function listTodo (request, reply) {
```

```
      return { data: [], totalCount: 0 } // [2]
    }
  })

  fastify.route({
    method: 'POST',
    url: '/',
    handler: async function createTodo (request, reply) {
      return { id: '123'} // [3]
    }
  })

  fastify.route({
    method: 'GET',
    url: '/:id',
    handler: async function readTodo (request, reply) {
      return {} // [4]
    }
  })

  fastify.route({
    method: 'PUT',
    url: '/:id',
    handler: async function updateTodo (request, reply) {
      reply.code(204) // [5]
    }
  })

  fastify.route({
    method: 'DELETE',
    url: '/:id',
    handler: async function deleteTodo (request, reply) {
      reply.code(204) // [6]
    }
  })

  fastify.route({
    method: 'POST',
    url: '/:id/:status',
    handler: async function changeStatus (request, reply) {
      reply.code(204) // [7]
    }
  })
}
```

Our module exports ([**1**]) is a Fastify plugin called `todoRoutes`. Inside it, we have defined six routes, five for basic CRUD operations and one additional action to flag tasks as done. Let's take a brief look at every one of them:

- `listTodo GET /`: Implements the **List** operation. It returns an array of to-do tasks and the total count of the elements ([**2**]).

- `createTodo POST /`: Implements the **Create** operation. It creates the task from the body data of `request` and returns `id` of the created element ([**3**]).

- `readTodo GET /:id`: Implements the **Read** operation. It returns the task that matches the `:id` parameter ([**4**]).

- `updateTodo PUT /:id`: Implements the **Update** operation. It updates the to-do item that matches the `:id` parameter, using the body data of `request` ([**5**]).

- `deleteTodo DELETE /:id`: Implements the **Delete** operation. It deletes the task matching the `:id` parameter ([**6**]).

- `changeStatus POST /:id/:status`: Implements a **custom action**. It marks a task as "done" or "undone" ([**7**]).

Note that we added a name to every handler function for clarity and as a best practice, since it helps to have better stack traces.

Now, let's look at how to use this plugin module inside our application.

Register routes

Solely declaring the routes plugin doesn't add any value to our application. Therefore, we need to register it before using it. Thankfully, we already have everything from the previous chapter to auto-register these routes. The following excerpt from `apps.js` shows the vital part:

```
// ...
fastify.register(AutoLoad, {
  dir: path.join(__dirname, 'routes'), // [1]
  indexPattern: /.*routes(\.js|\.cjs)$/i, // [2]
  ignorePattern: /.*\.js/,
  autoHooksPattern: /.*hooks(\.js|\.cjs)$/i,
  autoHooks: true,
  cascadeHooks: true,
  options: Object.assign({}, opts)
})
// ...
```

This code snippet uses a plugin called `@fastify/autoload` to automatically load routes and hooks from a specified directory.

We specified the `routes` folder ([1]) as the path where our routes are located, and then we defined the regular expression pattern ([2]) to identify the route files. Therefore, to make Fastify pick our previous `routes.js` file, we must save it in the `./routes/todos/routes.js` file.

You may be wondering why we added that `todos` subfolder to our path. `AutoLoad` has another neat behavior – it will automatically load all the subfolders of the specified path, using the folder name as the prefix path for the routes we define. Our handlers will be prefixed with the `todos` path when registered by the Fastify application. This feature helps us organize our code in subfolders without forcing us to define the prefix manually. Let's make a couple of calls to our application routes to make some concrete examples.

We need two terminals opened, the first to start the application and the second to make our calls using `curl`.

In the first terminal, go to the project root and type npm to start, as shown here:

```
$ npm start
{"level":30, "time":1679152261083, "pid":92481, "hostname": "dev.
local", "msg": "Server listening at http://127.0.0.1:3000"}
```

Now that the server is running, we can leave the first terminal open and go to the second one. We are ready to make the API calls:

```
$ curl http://127.0.0.1:3000/todos
{"data":[],"totalCount":0}%
$ curl http://127.0.0.1:3000/todos/1
{}%
```

In the preceding snippet, we can see we made two calls. In the first one, we successfully called the `listTodo` handler, while in the second call, we called `readTodo`.

Data source and model

Before implementing the handlers' logic, we need to look at data persistence. Since we registered the MongoDB plugin inside the application in *Chapter 6, Project Structure and Configuration Management*, we already have everything in place to save our to-do items to a real database.

Thanks to Fastify's plugin system, we can use the database client inside our route plugin, since the instance we receive as the first argument is decorated with the `mongo` property. Furthermore, we can assign the `'todos'` collection to a local variable and use it inside the route handlers:

```
'use strict'

module.exports = async function todoAutoHooks (fastify, _opts) {
  const todos = fastify.mongo.db.collection('todos')
```

```
    // ... rest of the route plugin implementation ...
}
```

We can now move on to defining our data model. Even if MongoDB is a schemaless database and we don't need to define anything upfront, we will outline a simple interface for a to-do task. It is important to remember that we don't need to add this code snippet to our application or database. We are showing it here just for clarity:

```
interface Todo {
    _id: ObjectId, // [1]
    id: ObjectId, // [2]
    title: string, // [3]
    done: boolean, // [4]
    createdAt: Date, // [5]
    modifiedAt: Date, // [6]
}
```

Let's take a look at the properties we just defined:

- _id ([1]) and id ([2]) have the same value. We add the id property not to expose any information about our database. The _id property is defined and used mainly by MongoDB servers.

- The title ([3]) property is user-editable and contains the to-do task.

- The done ([4]) property saves the task status. A task is completed when the value is true. Otherwise, a task is still in progress.

- createdAt ([5]) and modifiedAt ([6]) are automatically added by the application to track when the item was created and last modified.

Now that we have defined everything we need from the data source perspective, we can finally move on to implement the logic of the route handlers in the next section.

Implementing the routes

Until now, we implemented our handlers as dummy functions that don't do anything at all. This section will teach us how to save, retrieve, modify, and delete actual to-do tasks using MongoDB as the data source. For every subsection, we will examine only one handler, knowing that it will replace the same handler we already defined in ./routes/todos/routes.js.

> **Unique identifiers**
>
> This section contains several code snippets and commands to issue in the terminal. It is important to remember that the unique IDs we show here are different from the ones you will have when testing the routes. In fact, the IDs are generated when a task is created. Change the command snippets accordingly.

We will start with `createTodo` since having items saved on the database will help us implement and test the other handlers.

createTodo

As the name implies, this function allows users to create new tasks and save them to the database. The following code snippet defines a route that handles a POST request when a user hits the /todos/ path:

```
fastify.route({
    method: 'POST',
    url: '/',
    handler: async function createTodo (request, reply) { // [1]
      const _id = new this.mongo.ObjectId() // [2]
      const now = new Date() // [3]
      const createdAt = now
      const modifiedAt = now

      const newTodo = {
  _id,
      id: _id,
      ...request.body, // [4]
      done: false,
      createdAt,
      modifiedAt
    }

      await todos.insertOne(newTodo) // [5]
      reply.code(201) // [6]

      return { id: _id }
  }
})
```

When the route is invoked, the handler function ([1]) generates a new unique identifier ([2]) for the to-do item and sets the creation and modification dates ([3]) to the current time. The handler then constructs a new to-do object from the request body ([4]). The object is then inserted into the database using the `todos` collection we created at the beginning of the routes plugin ([5]). Finally, the function sends a response with a status code of 201 ([6]), indicating that the resource has been created and a body containing the ID of the newly created item.

At last, we can test our new route. As usual, we can use two terminal windows and `curl` to make calls, passing the body.

In the first terminal, run the server:

```
$ npm start
{"level":30, "time":1679152261083, "pid":92481, "hostname": "dev.
local", "msg": "Server listening at http://127.0.0.1:3000"}
```

Now in the second, we can use `curl` to perform the request:

```
$ curl -X POST http://localhost:3000/todos -H "Content-Type:
application/json" -d '{"title": "my first task"}'
{"id": "64172b029eb96017ce60493f"}%
```

We can see that the application returned `id` of the newly created item. Congratulations! You implemented your first working route!

In the following subsection, we will read the tasks list from the database!

listTodo

Now that our first item is saved to the database, let's implement the list route. It will allow us to list all the tasks with their total count.

We can start directly with the excerpt from `routes/todos/routes.js`:

```
fastify.route({
    method: 'GET',
    url: '/',
    handler: async function listTodo (request, reply) {
      const { skip, limit, title } = request.query // [1]
      const filter = title ? { title: new RegExp(title,
        'i') } : {} // [2]
      const data = await todos
        .find(filter, {
          limit,
          skip
        }) // [3]
        .toArray()

      const totalCount = await todos.countDocuments(filter)
      return { data, totalCount } // [4]
    }
})
```

Inside the `listTodo` function, the request object is used to extract query parameters ([1]), such as `skip`, `limit`, and `title`. The `title` parameter is used to create a regular expression filter to search for to-do items whose titles partially match the `title` parameter ([2]). If `title` is not provided, `filter` is an empty object, effectively returning all items.

The data variable is then populated with to-do items that match `filter`, by calling `todos.find()` and passing it as the parameter. In addition, the `limit` and `skip` query parameters are also passed to implement proper **pagination** ([3]). Since the MongoDB driver returns a cursor, we convert the result to an array using the `toArray()` method.

Pagination

Pagination is a technique used in database querying to limit the number of results returned by a query and retrieve only a specific subset of data at a time. When a query returns a large number of results, it can be difficult to display or process all of them at once. When working with lists of items, pagination allows users to access and process large amounts of data more manageably and efficiently. As a result, it improves the user experience and reduces the application and database load, leading to better performance and scalability.

The `totalCount` variable is calculated by calling `todos.countDocuments()` with the same `filter` object, so the API client can implement the pagination correctly. Finally, the handler function returns an object containing the `data` array and the `totalCount` number ([4]).

Again, we can now call the route using the two terminal instances and the `curl` binary, and we expect the response to have our first to-do item.

In the first terminal, run the server:

```
$ npm start
{"level":30, "time":1679152261083, "pid":92481, "hostname": "dev.
local", "msg": "Server listening at http://127.0.0.1:3000"}
```

Now in the second, we can use `curl` to perform the request:

```
$ curl http://127.0.0.1:3000/todos
{"data":[{"_id": "64172b029eb96017ce60493f", "title": "my
first task", "done":false, "id": "64172b029eb96017ce60493f",
"createdAt": "2023-03-19T15:32:18.314Z", "modifiedAt":
"2023-03-19T15:32:18.314Z"}], "totalCount":1}%
```

We can see that everything is working as expected, and `"my first task"` is the only item returned inside the `data` array. Also, `totalCount` is correctly number 1.

The next route we will implement allows us to query for one specific item.

readTodo

This RESTful route allows clients to retrieve a single to-do item from the database, based on its unique id identifier. The following excerpt illustrates the implementation of the handler function:

```
fastify.route({
    method: 'GET',
    url: '/:id', // [1]
    handler: async function readTodo (request, reply) {
      const todo = await todos.findOne(
        { _id: new this.mongo.ObjectId(request.params.id)
// [2]
      },
        { projection: { _id: 0 } }
      ) // [3]

      if (!todo) {
        reply.code(404)
        return { error: 'Todo not found' } // [4]
      }
      return todo // [5]
    }
  })
```

The /:id syntax in the url property ([1]) indicates that this route parameter will be replaced with a specific value when the client calls this route. In fact, the handler function first retrieves this id from the request.params object and creates a new ObjectId from it, using this.mongo. ObjectId() ([2]). It then uses the findOne method of the todos collection to retrieve the task with the matching _id. We exclude the _id field from the result, using the projection option, to not leak the database server we use ([3]). In fact, MongoDB is the only one that uses the _id field as the primary reference.

If a matching to-do item is found, it is returned as the response ([5]). Otherwise, the handler sets the HTTP status code to 404 and returns an error object, with a message saying the task was not found ([4]).

To test the route, we can use the usual process. From now on, we will omit the terminal that runs the server and only show the one we use to make our calls:

```
$ curl http://127.0.0.1:3000/todos/64172b029eb96017ce60493f
{"title": "my first task", "done":false, "id":
"64172b029eb96017ce60493f", "createdAt": "2023-03-19T15:32:18.314Z",
"modifiedAt": "2023-03-19T15:32:18.314Z"}%
```

Again, everything works as expected. We managed to pass the ID of the task we added to the database as the route parameter and received, as the response, the task titled "my first task".

So far, if a user makes a mistake in the title, there is no way to change it. This will be the next thing we will take care of.

updateTodo

The following code snippet adds a route that handles PUT requests to update a task already saved in the database:

```
fastify.route({
    method: 'PUT',
    url: '/:id', // [1]
    handler: async function updateTodo (request, reply) {
      const res = await todos.updateOne(
        { _id: new
        fastify.mongo.ObjectId(request.params.id) }, // [2]
        {
          $set: {
            ...request.body, // [3]
            modifiedAt: new Date()
          }
        }
      )
      if (res.modifiedCount === 0) { // [4]
        reply.code(404)
        return { error: 'Todo not found'}
      }

      reply.code(204) // [5]
    }
  })
```

We again use the :id parameter to identify which item the user wants to modify ([1]).

Inside the route handler, we use the MongoDB client updateOne() method to update the to-do item in the database. We are using the request.params.id property once more to create a filter object to match the task with the provided _id ([2]). Then, we use the $set operator to partially update the item with the new values from request.body. We also set the modifiedAt property to the current time ([3]).

After the update is completed, it checks the modifiedCount property of the result to see whether the update was successful ([4]). If no documents were modified, it returns a 404 error. If the update is successful, it produces a 204 status code to indicate it was completed successfully without returning a body ([5]).

After running the server in the usual way, we can test the route we just implemented using the terminal and `curl`:

```
$ curl -X PUT http://localhost:3000/todos/64172b029eb96017ce60493f
-H "Content-Type: application/json" -d '{"title": "my first task
updated"}'
```

This time, we pass the `-X` argument to `curl` to use the `PUT` HTTP method. Then, in the request body, we modify the title of our tasks and pass the task's unique ID as the route parameter. One thing that might create confusion is that the server hasn't returned a body, but looking at the return value of `updateTodo`, it shouldn't come as a surprise.

We can check whether the to-do item was updated correctly by calling the `readTodo` route:

```
$ curl http://127.0.0.1:3000/todos/64172b029eb96017ce60493f
{"title": "my first task updated", "done":false, "id":
"64172b029eb96017ce60493f", "createdAt": "2023-03-19T15:32:18.314Z",
"modifiedAt": "2023-03-19T17:41:09.520Z"}%
```

In the response, we can immediately see the updated title and the `modifiedAt` date, which is now different from `createdAt`, signaling that the item was updated.

Our application still lacks a delete functionality, so it is time to fix it. The following subsection will overcome this limitation.

deleteTodo

Following the RESTful conventions, the next code snippet defines a Fastify route that allows a user to delete a task, passing its unique `:id` as a request parameter:

```
fastify.route({
  method: 'DELETE',
  url: '/:id', // [1]
  handler: async function deleteTodo (request, reply) {
    const res = await todos.deleteOne({ _id: new
      fastify.mongo.ObjectId(request.params.id) }) // [2]
    if (res.deletedCount === 0) { // [3]
      reply.code(404)
      return { error: 'Todo not found'}
    }
    reply.code(204) // [4]
  }
})
```

After declaring the `DELETE` HTTP method, we pass the `:id` parameter as the route path to allow us to identify which item to delete ([1]).

Inside the `deleteTodo` function, we create the filter from the `request.params.id` property ([2]), and we pass it to the `todos` collection `deleteOne` method to delete tasks with that unique ID. After that call returns, we check whether the item was actually deleted from the database. If no documents were removed, the handler returns a `404` error ([3]). On the other hand, if the deletion is successful, we return an empty body with a `204` status code to indicate that the operation has completed successfully ([4]).

Testing the newly added route is simple, as always – we use the same terminal and `curl` setup we used for the previous routes.

After starting the server in one terminal, we run the subsequent command in the other:

```
$ curl -X DELETE http://localhost:3000/todos/64172b029eb96017ce60493f
$ curl http://127.0.0.1:3000/todos/64172b029eb96017ce60493f
{"error": "Todo not found"}%
```

Here, we make two different calls. The first one deletes the entity in the database and returns an empty response. The second, on the other hand, is to check whether the previous call deleted the resource. Since it returns a not found error, we are sure we deleted it.

No to-do list application would be complete without a way to mark tasks as "done" or move them back to "progress," and this is precisely the next thing we will add.

changeStatus

This is our first route that doesn't follow the CRUD principles. Instead, it is a custom logic that performs a specific operation on a single task. The following excerpt from `routes/todos/routes.js` shows a `POST` action that, upon invocation, marks a task as "done" or "not done," depending on its state. It is the first route to use two distinct request parameters:

```
fastify.route({
    method: 'POST',
    url: '/:id/:status', // [1]
    handler: async function changeStatus (request, reply) {
      const done = request.params.status === 'done' // [2]
      const res = await todos.updateOne(
        { _id: new
          fastify.mongo.ObjectId(request.params.id) },
        {
          $set: {
            done,
            modifiedAt: new Date()
          }
        }
      ) // [3]
      if (res.modifiedCount === 0) { // [4]
```

```
        reply.code(404)
        return { error: 'Todo not found'}
    }

    reply.code(204) // [5]
  }
})
```

Our route expects two parameters in the URL – `:id`, the unique identifier of the to-do item, and `:status`, which indicates whether the current task should be marked as "done" or "not done" ([1]).

The handler function first checks the value of the `status` parameter to determine the new value of the `done` property ([2]). It then uses the `updateOne()` method to update the `done` and `modifiedAt` properties of the item in the database ([3]). If the update is successful, the handler function returns a `204 No Content` response ([5]). On the other hand, if the item is not found, the handler function returns a `404 Not Found` response with an error message ([4]).

Before testing this route, we need at least one task in the database. If necessary, we can use the `createTodo` route to add it. Now, we can test the implementation using `curl`, as usual:

```
$ curl -X POST http://localhost:3000/todos/641826ecd5e0cccc313cda86/
done
$ curl http://localhost:3000/todos/641826ecd5e0cccc313cda86
{"id": "641826ecd5e0cccc313cda86", "title": "my first task",
"done":true, "createdAt": "2023-03-20T09:27:08.986Z", "modifiedAt":
"2023-03-20T09:27:32.902Z"}%
$ curl -X POST http://localhost:3000/todos/641826ecd5e0cccc313cda86/
undone
$ curl http://localhost:3000/todos/641826ecd5e0cccc313cda86
{"id": "641826ecd5e0cccc313cda86", "title": "my first task",
"done":false, "createdAt": "2023-03-20T09:27:08.986Z", "modifiedAt":
"2023-03-20T09:56:06.995Z"}
```

In the terminal output, we set the item's `done` property to `true`, passing `done` as the `:status` parameter of the request. We then call the `GET` single-item route to check whether the operation effectively changes the status. Then, to revert the process and mark the task as not yet done, we call the `done` route again, passing `undone` as the status request parameter. Finally, we check that everything works as expected and call again the `readTodo` handler.

This last route completes our to-do list application's basic functionalities. We are not done yet, though. In the next section, we will learn more about the security of our application and why our current implementation is insecure by design.

Securing the endpoints

So far, every route we declared doesn't perform any check on the input the user passes. This isn't good, and we, as developers, should always validate and sanitize the input of the APIs we expose. In our case, all the `createTodo` and `updateTodo` handlers are affected by this security issue. In fact, we take the `request.body` and pass it straight to the database.

First, to better understand the underlying issue, let's give an example of how a user can inject undesired information into our database with our current implementation:

```
$ curl -X POST http://localhost:3000/todos -H "Content-Type:
application/json" -d '{"title": "awesome task", "foo": "bar"}'
{"id": "6418214ad5e0cccc313cda85"}%
$ curl http://127.0.0.1:3000/todos/6418214ad5e0cccc313cda85
{"id": "6418214ad5e0cccc313cda85", "title": "awesome task", "foo":
"bar", "done":false, "createdAt": "2023-03-20T09:03:06.324Z",
"modifiedAt": "2023-03-20T09:03:06.324Z"}%
```

In the preceding terminal snippet, we issued two `curl` commands. In the first one, when creating an item, instead of passing only `title`, we also pass the `foo` property. Looking at the output returned, we can see that the command returned the ID of the created entity. Now, we can check what is saved in the database by calling the `readTodo` route. Unfortunately, we can see in the output that we also saved `"foo": "bar"` in the database. As previously mentioned, this is a security issue, and we should never allow users to write directly to the database.

There is another issue with the current implementation. We didn't attach any response serialization schema to our routes. While less critical from a security perspective, they are crucial regarding the throughput of our application. Letting Fastify know in advance the shape of the values we return from our routes helps it to serialize the response body faster. Therefore, we should always add all schemas when declaring a route.

In the upcoming sections, we will implement only one schema per type to make the exposition concise. You can find all schemas in the dedicated folder of the accompanying repository at `https://github.com/PacktPublishing/Accelerating-Server-Side-Development-with-Fastify/tree/main/Chapter%207/routes/todos/schemas`.

Loading route schemas

Before implementing the schemas, let's add a dedicated folder to organize our code base better. We can do it inside the `./routes/todos/` path. Moreover, we want to load them automatically from the `schemas` folder. To be able to do that, we need the following:

- A dedicated plugin inside the `schemas` folder
- A definition of the schemas we wish to use

- An autohooks plugin that will load everything automatically when the `todos` module is registered on the Fastify instance

We will discuss these in detail in the following subsections.

Schemas loader

Starting with the first item of the list we just discussed, we want to create a `./routes/todos/schemas/loader.js` file. We can check the content of the file in the following code snippet:

```
'use strict'

const fp = require('fastify-plugin')

module.exports = fp(async function schemaLoaderPlugin (fastify, opts)
{ // [1]
  fastify.addSchema(require('./list-query.json')) // [2]
  fastify.addSchema(require('./create-body.json'))
  fastify.addSchema(require('./create-response.json'))
  fastify.addSchema(require('./status-params.json'))
})
```

Let's break down this simple plugin:

- We defined a Fastify plugin named `schemaLoaderPlugin` that loads JSON schemas ([1])
- We called Fastify's `addSchema` method several times, passing the path of each JSON file as an argument ([2])

As we already know, every schema definition defines the structure and the validation rules of response bodies, parameters, and queries for different routes.

Now, we can start implementing the first body validation schema.

Validating the createTodo request body

The application will use this schema during task creation. We want to achieve two things with this schema:

1. Prevent users from adding unknown properties to the entity
2. Make the `title` property mandatory for every task

Let's take a look at the code of `create-body.json`:

```
{
  "type": "object",
  "$id": "schema:todo:create:body", // [1]
  "required": ["title"], // [2]
  "additionalProperties": false, // [3]
```

```
    "properties": {
      "title": {
        "type": "string" // [4]
      }
    }
  }
}
```

The schema is of the `object` type and, even if short in length, adds many constraints to the allowed inputs:

- `$id` is used to identify the schema uniquely across the whole application; it can be used to reference it in other parts of the code ([1]).

- The `required` keyword specifies that the `title` property is required for this schema. Any object that does not contain it will not be considered valid against this schema ([2]).

- The `additionalProperties` keyword is `false` ([3]), meaning that any properties not defined in the `"properties"` object will be considered invalid against this schema and discarded.

- The only property allowed is `title` of the `string` type ([4]). The validator will try to convert `title` to a string during the body validation phase.

Inside *Using the schemas* section, we will see how to attach this definition to the correct route. Now, we will move on and secure the request path parameters.

Validating the changeStatus request parameters

This time, we want to validate the request path parameters instead of a request body. This will allow us to be sure that the call contains the correct parameters with the correct type. The following `status-params.json` shows the implementation:

```
{
  "type": "object",
  "$id": "schema:todo:status:params", // [1]
  "required": ["id", "status"], // [2]
  "additionalProperties": false,
  "properties": {
    "id": {
      "type": "string" // [3]
    },
    "status": {
      "type": "string",
      "enum": ["done", "undone"] // [4]
    }
  }
}
```

Let's take a look at how this schema works:

- The $id field defines another unique identifier for this schema ([1]).

- In this case, we have two required parameters – id and status ([2]).

- The id property must be a string ([3]), while status is a string whose value can be "done" or "undone" ([4]). No other properties are allowed.

Next, we will explore how to validate the query parameters of a request using listTodos as an example.

Validating the listTodos request query

At this point, it should be clear that all schemas follow the same rules. A query schema is not an exception. However, in the list-query.json snippet, we will use schema reference for the first time:

```
{
  "type": "object",
  "$id": "schema:todo:list:query", // [1]
  "additionalProperties": false,
  "properties": {
    "title": {
      "type": "string" // [2]
    },
    "limit": {
      "$ref": "schema:limit#/properties/limit" // [3]
    },
    "skip": {
      "$ref": "schema:skip#/properties/skip"
    }
  }
}
```

We can now break down the snippet:

- As usual, the $id property gives the schema a unique identifier that can be referenced elsewhere in the code ([1]).

- The title property is of the string type, and it is optional ([2]). It can be filtered by the partial title of the to-do item. If not passed, the filter will be created empty.

- The limit property specifies the maximum number of items to return and is defined by referencing the schema schema:limit schema ([3]). The skip property is also defined by referencing schema schema:skip and is used for pagination purposes. These schemas are so general that they are shared throughout the project.

Now, it is time to take a look at the last schema type – the response schema.

Defining the createTodo response body

Defining a response body of a route adds two main benefits:

- It prevents us from leaking undesired information to clients

- It increases the throughput of the application, thanks to the faster serialization

`create-response.json` illustrates the implementation:

```
{
  "type": "object",
  "$id": "schema:todo:create:response", // [1]
  "required": ["id"], // [2]
  "additionalProperties": false,
  "properties": {
    "id": {
      "type": "string" // [3]
    }
  }
}
```

Let's examine the structure of this schema:

- Once again, `$id` is a unique identifier for this schema (**[1]**)

- The response object has one `required` (**[2]**) property, named `id`, of the `string` type (**[3]**)

This response schema ends the current section about schema definitions. Now, it is time to learn how to use and register those schemas.

Adding the Autohooks plugin

Once again, we can leverage the extensibility and the plugin system Fastify gives to developers. We can start by recalling from *Chapter 6* that we already registered a `@fastify/autoload` instance on our application. The following excerpt from the `app.js` file shows the relevant parts:

```
fastify.register(AutoLoad, {
  dir: path.join(__dirname, 'routes'),
  indexPattern: /.*routes(\.js|\.cjs)$/i,
  ignorePattern: /.*\.js/,
  autoHooksPattern: /.*hooks(\.js|\.cjs)$/i, // [1]
  autoHooks: true, // [2]
  cascadeHooks: true, // [3]
  options: Object.assign({}, opts)
})
```

For the purpose of this section, there are three properties that we care about:

- `autoHooksPattern` ([1]) is used to specify a regular expression pattern that matches the filenames of the hook files in the `routes` directory. These files will be automatically loaded and registered as hooks for the corresponding routes.

- `autoHooks` ([2]) enables the automatic loading of those hook files.

- `cascadeHooks` ([3]) ensures that the hooks are executed in the correct order.

After this brief reminder, we can move on to implementing our autohook plugin.

Implementing the Autohook plugin

We learned from `autoHooksPattern` in the previous section that we can put our plugin inside a file named `autohooks.js` in the `./routes/todos` directory, and it will be automatically registered by `@fastify/autoload`. The following snippet contains the content of the plugin:

```
'use strict'

const fp = require('fastify-plugin')
const schemas = require('./schemas/loader') // [1]

module.exports = fp(async function todoAutoHooks (fastify, opts) {
  fastify.register(schemas) // [2]
})
```

We start importing the schema loader plugin we defined in a previous section ([1]). Then, inside the plugin body, we register it ([2]). This one line is enough to make the loaded schemas available in the application. In fact, the plugin attaches them to the Fastify instances to make them easily accessible.

Finally, we can use these schemas inside our route definitions, which we will do in the next section.

Using the schemas

Now, we have everything in place to secure our routes and make the application's throughput ludicrously fast.

This section will only show you how to do it for one route handler. You will find the complete code in the book's repository at `https://github.com/PacktPublishing/Accelerating-Server-Side-Development-with-Fastify/tree/main/Chapter%207`, and you are encouraged to experiment with other routes too.

The following code snippet attaches the schemas to the route definition:

```
fastify.route({
  method: 'POST',
```

```
    url: '/',
    schema: {
      body: fastify.getSchema('schema:todo:create:body'), // [1]
      response: {
        201: fastify.getSchema('schema:todo:create:response') // [2]
      }
    },
    handler: async function createTodo (request, reply) {
      // ...omitted for brevity
    }
  })
```

We are adding a `schema` property to the route definition. It contains an object with two fields:

- The `body` property of the `schema` option specifies the JSON schema that the request body must validate against ([1]). Here, we use `fastify.getSchema` (`'schema:todo:create:body'`), which retrieves the JSON schema for the request body from the schemas collection, using the ID we specified in the declaration.

- The `response` property of the `schema` option specifies the JSON schema for the response to the client ([2]). It is set to an object with a single key, `201`, which specifies the JSON schema for a successful creation response, since it is the code we used inside the handler. Again, we use `fastify.getSchema('schema:todo:create:response')` to retrieve the JSON schema for the response from the schemas collection.

If we now try to pass an unknown property, the schema validator will strip away from the body. Let's experiment with it using the terminal and `curl`:

```
$ curl -X POST http://localhost:3000/todos -H "Content-Type:
application/json" -d '{"title": "awesome task", "foo": "bar"}'
{"id":"6418671d625e3ba28a056013"}%
$ curl http://localhost:3000/todos/6418671d625e3ba28a056013
{"id":"6418671d625e3ba28a056013","title":"awesome task","done":fa
lse,"createdAt":"2023-03-20T14:01:01.658Z","modifiedAt":"2023-03-
20T14:01:01.658Z"}%
```

We pass the `foo` property inside our body, and the API returns a successful response, with the unique `id` of the task saved in the database. The second call checks that the validator works as expected. The `foo` field isn't present in the resource, and therefore, it means that our API is now secure.

This almost completes our deep dive into RESTful API development with Fastify. However, there is one more important thing that can make our code base more maintainable, which we need to mention before moving on.

Don't repeat yourself

Defining the application logic inside routes is fine for simple applications such as the one in our example. In a real-world scenario, though, when we need to use our logic across an application in multiple routes, it would be nice to define that logic only once and reuse it in different places. So, once more, Fastify has us covered.

We can expand our `autohooks.cjs` plugin by adding what is commonly known as a data source. In the following snippet, we expand the previous plugin, adding the needed code, although, for the brevity of the exposition, we are showing only the function for the `createTodo` handler; you can find the whole implementation inside the book's code repository:

```
'use strict'
const fp = require('fastify-plugin')
const schemas = require('./schemas/loader')
module.exports = fp(async function todoAutoHooks (fastify, opts) { //
[1]
  const todos = fastify.mongo.db.collection('todos') // [2]

  fastify.register(schemas)

  fastify.decorate('mongoDataSource', { // [3]
    // ...
    async createTodo ({ title }) { // [4]
      const _id = new fastify.mongo.ObjectId()
      const now = new Date()
      const { insertedId } = await todos.insertOne({
        _id,
        title,
        done: false,
        id: _id,
        createdAt: now,
        modifiedAt: now
      })
      return insertedId
    },
    // ...
})
```

Let's break down the implementation:

- We wrap our plugin with `fastify-plugin` to expose the data source to other plugin scopes ([1]).

- Since we will not access the MongoDB collection from the routes anymore, we moved its reference here ([2]).

- We decorate the Fastify instance with the `mongoDataSource` object ([**3**]) which has several methods, including `createTodo`.

- We moved the item creation logic that was inside the route handler here ([**4**]). The function returns `insertedId`, which we can use to populate the body to return to the clients.

Now, we must update our `createTodo` route handler to take advantage of the newly added code. Let's do it in the `routes/todos/routes.js` code excerpt:

```
fastify.route({
    method: 'POST',
    url: '/',
    schema: {
      body: fastify.getSchema('schema:todo:create:body'),
      response: {
        201:
          fastify.getSchema('schema:todo:create:response')
      }
    },
    handler: async function createTodo (request, reply) {
      const insertedId = await
      this.mongoDataSource.createTodo(request.body) // [1]
      reply.code(201)
      return { id: insertedId } // [2]
    }
```

Our handler body is a one-liner. Its new duty is to take `request.body` ([**1**]) and to pass it to the `createTodo` data source method. After that call returns, it takes the returned unique ID and forwards it to the client ([**2**]). Even in this simple example, it should be clear how powerful this feature is. We can use it to make our code reusable from every application part.

This final section covered everything we need to know to develop a simple yet complete application using Fastify.

Summary

This chapter taught us step by step how to implement a RESTful API in Fastify. First, we used the powerful plugin system to encapsulate our route definitions. Then, we secured our routes and database accesses using schema definitions. Finally, we moved the application logic inside a dedicated plugin using decorators. This allowed us to follow the DRY pattern and make our application more maintainable.

The following chapter will look at user management, sessions, and file uploads to extend the application's capabilities even more.

Authentication, Authorization, and File Handling

In this chapter, we will continue evolving our application, mainly covering two distinct topics: user authentication and file handling. First, we will implement a reusable JWT authentication plugin that will allow us to manage users, authentication, and sessions. It will also act as an authorization layer, protecting our application's endpoints from unauthorized access. We will also see how decorators can expose the authenticated user's data inside the route handlers. Then, moving on to file handling, we will develop a dedicated plugin enabling users to import and export their to-do tasks in CSV format.

In this chapter, we will learn about the following:

- Authentication and authorization flow
- Building the authentication layer
- Adding the authorization layer
- Managing uploads and downloads

Technical requirements

To follow along with this chapter, you will need these technical requirements mentioned in the previous chapters:

- A working Node.js 18 installation (`https://nodejs.org/`)
- VS Code IDE (`https://code.visualstudio.com/`)
- An active Docker installation (`https://docs.docker.com/get-docker/`)
- A Git repository is recommended but not mandatory (`https://git-scm.com/`)
- A terminal application

Once more, the code of the project can be found on GitHub at `https://github.com/PacktPublishing/Accelerating-Server-Side-Development-with-Fastify/tree/main/Chapter%208`.

Finally, it's time to start our exploration. In the next section, we will take a deep dive into the authentication flow in Fastify, understanding all the pieces we need to implement a complete solution.

Authentication and authorization flow

Authentication and authorization are usually challenging topics. Based on use cases, specific strategies may or may not be feasible. For this project, we will implement the authentication layer via **JSON Web Tokens**, commonly known as **JWTs**.

> JWT
>
> This is a widely used standard for token-based authentication for web and mobile applications. It is an open standard that allows information to be transmitted securely between the client and the server. Every token has three parts. First, the header contains information about the type of token and the cryptographic algorithms used to sign and encrypt the token. Then, the payload includes any metadata about the user. Finally, the signature is used to verify the token's authenticity and ensure it has not been tampered with.

Before looking at the implementation in Fastify, let's briefly explore how this **authentication** works. First, the API needs to expose an endpoint for the registration. This route will enable users to create new accounts on the service. After the account is created correctly, the user can perform authenticated operations against the server. We can break them down into seven steps:

1. To initiate the authentication process, the user provides their username and password to the server via a specific endpoint.

2. The server verifies the credentials and, if valid, creates a JWT containing the user's metadata using the shared secret.

3. The server returns the token to the client.

4. The client stores the JWT in a secure location. Inside the browser, it is usually local storage or a cookie.

5. On subsequent requests to the server, the client sends the JWT in the `Authorization` header of each HTTP request.

6. The server verifies the token by checking the signature, and if the signature is valid, it extracts the user's metadata from the payload.

7. The server uses the user ID to look up the user in the database.

From here on, the request is handled by the **authorization** layer. First, it must check whether the current user has the necessary permissions to perform the action or access the specified resource. Then, based on the result of the check operation, the server can answer with the resource or an HTTP Unauthorized error. They are many standardized ways of implementing authorization. In this book, we will implement our simple solution from scratch for exposition purposes.

> **Authentication versus authorization**
>
> Even if these terms are often used together, they express two completely different concepts. Authentication describes *who* is allowed to access the service. On the other hand, authorization defines *what* actions can be performed by the user once authenticated.

The authorization and authentication layers are crucial to building secure web applications. Controlling access to resources helps to prevent unauthorized access and protect sensitive data from potential attacks or breaches.

In the next section, we will start from where we left the code in *Chapter 7*, implementing a new application-level plugin for authentication.

Building the authentication layer

Since we need to add new non-trivial functionality to our application, we need to implement mainly two pieces:

- An authentication plugin to generate tokens, check the incoming requests, and revoke old or not used tokens

- A bunch of new routes to handle the registration, authentication, and life cycle of the tokens

Before jumping directly into the code, we need to add one last note. In this chapter's code snippets, we will use a new data source called userDataSource ([1]). Since it exposes only createUser ([3]) and readUser ([2]) methods and the implementation is trivial, we will not show it in this book. However, the complete code is in the ./routes/auth/autohooks.js file inside the GitHub repository.

Since we must implement both, we can add the authentication plugin first.

Authentication plugin

First, create the ./plugins/auth.js file inside the project's root folder. The auth.js code snippet shows the implementation of the plugin:

```
const fp = require('fastify-plugin')
const fastifyJwt = require('@fastify/jwt') // [1]
```

```
module.exports = fp(async function authenticationPlugin (fastify,
opts) {
  const revokedTokens = new Map() // [2]

  fastify.register(fastifyJwt, { // [3]
    secret: fastify.secrets.JWT_SECRET,
    trusted: function isTrusted (request, decodedToken) {
      return !revokedTokens.has(decodedToken.jti)
    }
  })

  fastify.decorate('authenticate', async function
  authenticate (request, reply) { // [4]
    try {
      await request.jwtVerify() // [5]
    } catch (err) {
      reply.send(err)
    }
  })

  fastify.decorateRequest('revokeToken', function () { //
  [6]
    revokedTokens.set(this.user.jti, true)
  })

  fastify.decorateRequest('generateToken', async function
  () { // [7]
    const token = await fastify.jwt.sign({
      id: String(this.user._id),
      username: this.user.username
    }, {
      jti: String(Date.now()),
      expiresIn: fastify.secrets.JWT_EXPIRE_IN
    })

    return token
  })
}, {
  name: 'authentication-plugin',
  dependencies: ['application-config']
})
```

We create and export a Fastify plugin that provides authentication functionalities via decorators and the JWT library. But first, let's take a look at the implementation details:

- We require the official `@fastify/jwt` package ([1]). It handles low-level primitives around the tokens and lets us focus only on the logic we need inside our application.

- Generally speaking, keeping a trace of the invalidated tokens is always a good idea. `revokedTokens` creates a `Map` instance ([2]) to keep track of them. Later, we will use it to ban invalid tokens.

- We register the `@fastify/jwt` plugin on the Fastify instance ([3]), passing the `JWT_SECRET` environment variable and `isTrusted` function that checks whether a token is trusted. In a subsequent section, we will add `JWT_SECRET` to our server's configuration to ensure its presence after the boot.

- We decorate the Fastify instance with the `authenticate` function to verify that the client's token is valid before allowing access to protected routes. The `request.jwtVerify()` ([5]) method comes from `@fastify/jwt`. If errors are thrown during the verification, the function replies to the client with the error. Otherwise, the `request.user` property will be populated with the current user.

- The `revokeToken` function is added to the Fastify instance ([6]). It adds a token to the map of invalid tokens. We use the `jti` property as the invalidation key.

- The `generateToken` function creates a new token from user data ([7]). Then, we decorate a request with this function to access its context through `this` reference. The `fastify.jwt.sign` method is once more provided by the `@fastify/jwt` library.

Thanks to the project setup from the previous chapters, this plugin will be automatically registered to the main Fastify instance inside `./apps.js` during the boot phase.

We can leave this file as is for now since we will start using the decorated methods inside our application in a dedicated section. Now, it is time to add the authentication layer routes, and we will do it in the following subsection.

Authentication routes

The time has come to implement a way for the users to interact with our authentication layer. The `./routes/auth` folder structure mimics the `todos` module we explored in *Chapter 7*. It contains `schemas`, `autohooks.js`, and `routes.js`. We will look only at `routes.js` in the book for brevity's sake. The rest of the code is straightforward and can be found in the GitHub repository at `https://github.com/PacktPublishing/Accelerating-Server-Side-Development-with-Fastify/tree/main/Chapter%208`.

Since the code of `./routes/auth/routes.js` is pretty long, we will split it into single snippets, one per route definition. But first, to get a general idea of the plugin, the following snippet contains the whole code while omitting the implementations:

```
const fp = require('fastify-plugin')

const generateHash = require('./generate-hash') // [1]

module.exports.prefixOverride = '' // [2]
module.exports = fp(
  async function applicationAuth (fastify, opts) {
    fastify.post('/register', {
      // ... implementation omitted
    })

    fastify.post('/authenticate', {
      // ... implementation omitted
    })

    fastify.get('/me', {
      // ... implementation omitted
    })

    fastify.post('/refresh', {
      // ... implementation omitted
    })

    fastify.post('/logout', {
      // ... implementation omitted
    })

    async function refreshHandler (request, reply) {
      // ... implementation omitted
    }
  }, {
    name: 'auth-routes',
    dependencies: ['authentication-plugin'], // [3]
    encapsulate: true
  })
```

We start by requiring the `generate-hash.js` local module ([1]). We don't want to save users' passwords in plain text, so we use this module to generate a hash and a salt to store in the database. Again, you can find the implementation in the GitHub repository. Next, since we want to expose the five routes declared in the body of the plugin directly on the root path, we set the `prefixOverride` property to an empty string and exported it ([2]). Since we are inside the `./routes/auth` subfolder, `@fastify/autoload` would instead mount the routes to the `/auth/` path. Furthermore, since inside our route declarations, we rely on methods decorated in `authentication-plugin`, we add it to the `dependencies` array ([3]). Finally, we want to override the default behavior of `fastify-plugin` to isolate this plugin's code, and therefore, we pass `true` to the `encapsulate` options.

This wraps up the general overview. Next, we can examine the `register` route.

Register route

This route allows new users to register on our platform. Let's explore the implementation by looking at the following snippet:

```
fastify.post('/register', { // [1.1]
  schema: {
    body: fastify.getSchema('schema:auth:register') //
    [1.2]
  },
  handler: async function registerHandler (request, reply)
  {
    const existingUser = await
    this.usersDataSource.readUser(request.body.username) //
    [1.3]
    if (existingUser) { // [1.4]
      const err = new Error('User already registered')
      err.statusCode = 409
      throw err
    }

    const { hash, salt } = await
      generateHash(request.body.password) // [1.5]

    try {
      const newUserId = await
        this.usersDataSource.createUser({ // [1.6]
        username: request.body.username,
        salt,
        hash
      })
      request.log.info({ userId: newUserId },
        'User registered')
```

```
      reply.code(201)
      return { registered: true } // [1.7]
    } catch (error) {
      request.log.error(error, 'Failed to register user')
      reply.code(500)
      return { registered: false } // [1.8]
    }
  }
})
```

Let's break down the execution of the preceding code snippet:

- First, `fastify.post` is used to declare a new route for the HTTP POST method with the `/register` path ([**1.1**]).

- We specify the request body schema using `fastify.getSchema` ([**1.2**]). We will not see this schema implementation in the book, but it can be found in the GitHub repository as usual.

- Moving to the handler function details, we use `request.body.username` to check whether the user is already registered to the application ([**1.3**]). If so, we throw a `409` HTTP error ([**1.4**]). Otherwise, `request.body.password` is passed to the `generateHash` function to create a hash and a salt from it ([**1.5**]).

- Then, we use these variables and `request.body.username` to insert the new user in the DB ([**1.6**]).

- If no errors are thrown during this creation process, the handler replies with a `201` HTTP code and a `{ registered: true }` body ([**1.7**]). On the other hand, if there are errors, the reply contains a `500` HTTP code and a `{ registered: false }` body ([**1.8**]).

The following section will examine how the users authenticate with our platform.

Authenticate route

The next route on the list is the POST `/authenticate` route. It allows registered users to generate a new JWT token using their password. The following snippet shows the implementation:

```
fastify.post('/authenticate', {
  schema: { // [2.1]
    body: fastify.getSchema('schema:auth:register'),
    response: {
      200: fastify.getSchema('schema:auth:token')
    }
  },
  handler: async function authenticateHandler (request,
  reply) {
```

```
        const user = await
          this.usersDataSource.readUser(request.body.username)
          // [2.2]

        if (!user) { // [2.3]
          // if we return 404, an attacker can use this to find
             out which users are registered
          const err = new Error('Wrong credentials provided')
          err.statusCode = 401
          throw err
        }

        const { hash } = await
        generateHash(request.body.password, user.salt) // [2.4]
        if (hash !== user.hash) { // [2.5]
          const err = new Error('Wrong credentials provided')
          err.statusCode = 401
          throw err
        }

        request.user = user // [2.6]
        return refreshHandler(request, reply)// [2.7]
    }
})
```

Let's break down the code execution:

- Once more, we use the auth schemas we declared in the dedicated folder ([**2.1**]) to secure and speed the route's body and response payloads.

- Then, inside the handler function, we read the user's data from the database using the `request.body.username` property ([**2.2**]).

- If no user is found in the system, we return `401` instead of `404`, with the **Wrong credentials provided** message, to prevent attackers from discovering which users are registered ([**2.3**]).

- We are now able to use the `user.salt` property we got from the database to generate a new hash ([**2.4**]). The generated `hash` is then compared with the hash stored in the data source during the user registration.

- If they do not match, the function throws the same `401` error using the `throw` statement ([**2.5**]).

- On the other hand, if the check is successful, the now authenticated user is attached to the request object for further processing ([**2.6**]).

- Finally, the handler invokes the `refreshHandler` function, passing `request` and `reply` as arguments ([**2.7**]).

We will see the `refreshHandler` implementation in the following section, where we look at the `/refresh` route.

Refresh route

Once authenticated, the `refresh` route allows our users to generate more tokens without providing their usernames and passwords. Since we already saw that we are using the same logic inside the `authenticate` route, we moved this route handler to a separate function. The following code block shows these details:

```
fastify.post('/refresh', {
  onRequest: fastify.authenticate, // [3.1]
  schema: {
    headers: fastify.getSchema('schema:auth:token-header'),
    response: {
      200: fastify.getSchema('schema:auth:token')
    }
  },
  handler: refreshHandler // [3.2]
})

async function refreshHandler (request, reply) {
  const token = await request.generateToken() // [3.3]
  return { token }
}
```

This route is the first one protected by the authentication layer. To enforce it, we use `fastify.authenticate onRequest` hook, which we created in the *Authentication plugin* section ([**3.1**]). The route handler function is `refreshHandler` ([**3.2**]), which generates a new JWT token and returns it as the response. Finally, the handler calls the `generateToken` method decorated onto the request object by the authentication plugin ([**3.3**]) and then returns its value to the client. The route is authenticated because we generate the new token from the request called by already authorized users.

The time has come to look at how we invalidate user tokens, and we will do precisely that in the next section.

Logout route

Until now, we didn't use the `revokedTokens` map and `revokeToken` request method we created in the *Authentication plugin* section. However, the logout implementation relies on them. Let's jump into the code:

```
fastify.post('/logout', {
  onRequest: fastify.authenticate, // [4.1]
  handler: async function logoutHandler (request, reply) {
```

```
      request.revokeToken() // [4.2]
      reply.code(204) // [4.3]
    }
  })
```

Since we want only authenticated users to invalidate their tokens, the /logout route is once more protected by the authentication hook ([**4.1**]). Assuming the request authentication succeeds, the handler function revokes the current token calling the request.revokeToken method ([**4.2**]), which is attached to the request object by the authentication plugin we developed previously. This call adds the token to the revokedTokens map used internally by the @fastify/jwt plugin to determine invalid entries. The token revocation process ensures that the token cannot be used again for authentication, even if an attacker manages to obtain it. Finally, the handler sends an empty 204 response to the client, indicating a successful logout ([**4.3**]).

This completes this section about the authentication routes. In the following one, we will implement our authorization layer.

Adding the authorization layer

Now that we can have all authentication pieces in place, we can finally move on to implementing the authorization layer of our application. To adequately protect our endpoints, we need to do two main things to the ./routes/todos module from *Chapter 7*:

- Add the authentication layer to ./routes/todos/routes.js
- Update the to-do data source inside ./routes/todos/autohook.js

Fortunately, we need only a one-liner change to implement the first point. On the other hand, the second point is more complex. We will examine both in the following subsections.

Adding the authentication layer

Let's start with the simpler task. As we already said, this is a fast addition to the *Chapter 7* code that we can see in the following snippet:

```
module.exports = async function todoRoutes (fastify, _opts) {
  fastify.addHook('onRequest', fastify.authenticate) // [1]
  // omitted route implementations from chapter 7
}
```

To protect our to-do routes, we add the onRequest fastify.authenticate hook ([**1**]), which we previously used for authentication routes. This hook will check whether the incoming request has the authentication HTTP header, and after validating it, it will add the user information object to the request.

Updating the to-do data source

Since our application deals only with one type of entity, our authorization layer is straightforward to implement. The idea is to prevent users from accessing and modifying tasks that belong to other users. Until this point, we could see our application as a single-user application:

- Every task we created doesn't have any reference to the user that created it

- Every Read, Update, and Delete operation can be executed on every item by every user

As we already said, the correct place to fix these issues is the mongoDataSource decorator we implemented in *Chapter 7*. Since we now have two data sources, one for the users and the other for the to-do items, we will rename mongoDataSource to todosDataSource to reflect its duties better. Because we need to change all methods to add a proper authorization layer, the code snippet would get too long. Instead of showing its entirety here, the following snippet shows changes only for listTodos and createTodos. All changes can be found in the ./routes/todos/ autohooks.js file inside the GitHub repository of this chapter:

```js
// ... omitted for brevity
module.exports = fp(async function todoAutoHooks (fastify, opts) {
  // ... omitted for brevity
  fastify.decorateRequest('todosDataSource', null) // [1]
  fastify.addHook('onRequest', async (request, reply) => {
  // [2]
    request.todosDataSource = { // [3]
      // ... omitted for brevity
      async listTodos ({
        filter = {},
        projection = {},
        skip = 0,
        limit = 50
      } = {}) {
        if (filter.title) {
          filter.title = new RegExp(filter.title, 'i')
        } else {
          delete filter.title
        }
        filter.userId = request.user.id // [4]

        const data = todos
          .find(filter, {
            projection: { ...projection, _id: 0 },
            limit,
            skip
          }).toArray()

        return data
```

```
      },
      async createTodo ({ title }) {
        const _id = new fastify.mongo.ObjectId()
        const now = new Date()
        const userId = request.user.id // [5]
        const { insertedId } = await todos.insertOne({
          _id,
          userId,
          title,
          done: false,
          id: _id,
          createdAt: now,
          modifiedAt: now
        })
        return insertedId
      }
      // ... omitted for brevity
    }
  })
}
```

Instead of decorating the Fastify instance as we did initially in *Chapter 7*, we are now moving the logic inside the `request` object. This change allows easy access to the `user` object that our authentication layer attaches to the request. Later, we will use this data across all `todosDataSource` methods.

Let's take a closer look at the code:

- First, we decorate the request with `todosDataSource`, setting its value to `null` ([1]). We do this to trigger a speed optimization: making the application aware of the existence of the `todosDataSource` property at the beginning of the request life cycle will make its creation faster.

- Then, we add the `onRequest` hook ([2]), which will be called after `fastify.authentication` has already added the user data.

- Inside the hook, a new object containing the data source implementations is assigned to the `todosDataSource` property on the request ([3]).

- Next, `listTodos` now uses `request.user.id` as a filter field ([4]) to return only the data that belongs to the current user.

- To make this filter work, we must add the `userId` property to the newly created tasks ([5]).

As we said, we omit the other methods for brevity, but they follow the same pattern using `userId` as a filter. Again, the complete code is present in the GitHub repository.

We just completed our authentication and authorization layers. The next section will show how to handle file uploads and downloads inside authentication-protected endpoints.

Managing uploads and downloads

We need to add two more functionalities to our application, and we will do it by developing a dedicated Fastify plugin. The first will allow our users to upload CSV files to create to-do tasks in bulk. We will rely on two external dependencies to do it:

- `@fastify/multipart` for file uploads

- `csv-parse` for CSV parsing

The second plugin will expose an endpoint to download items as a CSV file. Again, we need the external `csv-stringify` library to serialize objects and create the document.

While we will split the code into two snippets in the book, the complete code can be found in `./routes/todos/files/routes.js`. Let's explore the following snippet, which contains the file upload and items bulk creation logic:

```
const fastifyMultipart = require('@fastify/multipart')
const { parse: csvParse } = require('csv-parse')
// ... omitted for brevity
  await fastify.register(fastifyMultipart, { // [1]
    attachFieldsToBody: 'keyValues',
    sharedSchemaId: 'schema:todo:import:file', // [2]
    async onFile (part) { // [3]
      const lines = []
      const stream = part.file.pipe(csvParse({ // [4]
        bom: true,
        skip_empty_lines: true,
        trim: true,
        columns: true
      }))
      for await (const line of stream) { // [5]
        lines.push({
          title: line.title,
          done: line.done === 'true'
        })
      }
      part.value = lines // [6]
    },
    limits: {
      fieldNameSize: 50,
      fieldSize: 100,
      fields: 10,
      fileSize: 1_000_000
      files: 1
    }
```

```
  })
  fastify.route({
    method: 'POST',
    url: '/import',
    handler: async function listTodo (request, reply) {
      const inserted = await
        request.todosDataSource.createTodos(
          request.body.todoListFile) // [7]
      reply.code(201)
      return inserted
    }
  })
// ... omitted for brevity
```

Let's go through the code execution:

- First, we register the `@fastify/multipart` plugin to the Fastify instance ([1]).

- To be able to access the content of the uploaded file directly from `request.body`, we pass the `attachFieldsToBody` and `sharedSchemaId` options ([2]).

- Next, we specify the `onFile` option property ([3]) to handle incoming streams. This function will be called for every file in the incoming request.

- Then, we use the `csvParse` library to transform the file into a stream of lines ([4]).

- A `for await` loop iterates over each parsed line ([5]) and transforms the data from each line, adding it to the `lines` array, and we assign the array to `part.value` ([6]).

- Finally, thanks to the options we passed to `@fastify/multipart`, we can access the `lines` array directly from `request.body.todoListFile` and use it as the argument for the `createTodos` method ([7]).

Once more, we are omitting the `createTodos` implementation, which can be found in the GitHub repository.

We can now move on to the endpoint for tasks exporting. The following snippet shows the implementation:

```
fastify.route({
    method: 'GET',
    url: '/export',
    schema: {
      querystring:
        fastify.getSchema('schema:todo:list:export')
    },
    handler: async function listTodo (request, reply) {
      const { title } = request.query
      const cursor = await
```

```
        request.todosDataSource.listTodos({ // [1]
        filter: { title },
        skip: 0,
        limit: undefined,
        asStream: true // [2]
      })

      reply.header('Content-Disposition', 'attachment;
        filename="todo-list.csv"')
      reply.type('text/csv') //[3]
      return cursor.pipe(csvStringify({ // [4]
        quoted_string: true,
        header: true,
        columns: ['title', 'done', 'createdAt',
          'updatedAt', 'id'],
        cast: {
          boolean: (value) => value ? 'true' : 'false',
          date: (value) => value.toISOString()
        }
      }))
    }
  })
```

We call the `listTodos` method of the `request.todosDataSource` ([1]) object to retrieve the list of to-do tasks that match the optional `title` parameter. If no title is passed, then the method will return all items. Moreover, thanks to our authentication layer, we know they will be automatically filtered based on the current user. The `asStream` option is set to `true` to handle cases where the data could be massive ([2]). The `Content-Disposition` header is set to specify that the response is an attachment with a filename of `todo-list.csv` ([3]). Finally, the cursor stream is piped to the `csvStringify` function to convert the data to a CSV file, which is then returned as the response body ([4]).

With this last section, we significantly increased the capabilities of our application, allowing users to import and export their tasks efficiently.

Summary

In this chapter, we added an authentication layer to ensure that only registered users can perform actions on the to-do items. Moreover, thanks to the modest authorization layer, we ensured that users could only access tasks they created. Finally, we showed how simple upload and download capabilities are to implement using a real-world example of bulk imports and exports.

In the next chapter, we will learn how to make our application reliable in production. We will use the tools that Fastify integrates to test our endpoints thoroughly. We want to prevent introducing any disruptions for our users because of lousy code pushed to production.

9

Application Testing

Proper testing is a primary, essential aspect that will make an application reliable for years to come. Fastify comes with several integrated tools specifically developed for making the testing experience as slick as possible. This means that writing an application test will not be frustrating and slow to run. Nevertheless, Fastify is test runner-agnostic; it perfectly integrates with the runner of your choice.

In this chapter, you will learn how to use integrated methods and run tests in parallel without having to spin an actual HTTP server.

The learning path we will cover in this chapter is as follows:

- Writing good tests
- Testing the Fastify application
- Dealing with complex tests
- Speeding up the test suite
- Where tests should run

Technical requirements

To go through this chapter, you will need the following:

- A working Node.js 18 installation
- The VSCode IDE from `https://code.visualstudio.com/`
- A Docker installation
- A public GitHub repository
- A working command shell

All the snippets in this chapter are on GitHub at `https://github.com/PacktPublishing/Accelerating-Server-Side-Development-with-Fastify/tree/main/Chapter%209`.

Writing good tests

Implementing tests for an application is an investment to free you up from the stress that a new project could give you. Thanks to these tests, you can be (pretty) sure of implementing new features or fixing bugs without breaking the existing software. Unfortunately, writing tests often gets sacrificed whenever a project is running out of time. This is also an activity that many consider to be boring and frustrating, due to the many blockers you may face by testing an HTTP server, such as the server port already in use. So, the tests are often written quickly and without the required attention.

We can say that an application's test suite is successful under the following conditions:

- It's automated – you just need to click a button to get many hours' worth of work done

- It's easy to maintain – adding a new test must be easy, and it should not take more time than developing the feature itself

- Running the tests must not take a lot of time

- Every developer can run the tests locally on their PC

Fastify's team knows this, and it has worked to give you the right tooling to support you on these focal tasks for a project's success.

We will explore the world of testing to learn how to write proper tests for the application we have developed so far, since *Chapter 6*.

So, let's jump into a quick introduction to testing.

What tests need to be written?

There are tons of books about tests and how to write them, and this chapter does not aim to replace those sources. This book offers you a practical approach based on our experience, covering the most common use cases you will face during your development.

You will need to distinguish these main test categories:

- **Unit tests**: Check that a small piece of software works as expected. Unit tests are mainly used to verify utilities or plugins. For example, testing a utility that sanitizes the user's input is a unit test.

- **Integration tests**: These aim to ensure that all the pieces of your application work together. Usually, integration tests require an additional setup to run correctly because they rely on external services, such as a database or a third-party API. For example, making an HTTP request to test that a row has been inserted into a database is an integration test.

- **Functional tests**: these tests track business requirements and real-world scenarios replicating the application usage. For example, a functional test checks whether the user can register and log in successfully to our system.

- **Regression tests**: Every solved bug in your application must be traced with a test that fails before the fix is applied to the code. For example, every bug fix should accompany a regression test associated with whatever the implementation requires, such as a running database or unusual API usage by the client.

Regardless of the framework used, every application needs this minimal set of tests to provide valuable support to the project.

The test pyramid

To deepen the tests, you should read these articles: `https://martinfowler.com/ bliki/TestPyramid.html` and `https://martinfowler.com/articles/ practical-test-pyramid.html`. These articles explain in detail how tests impact your application life cycle. The former analyzes that there is a monetary cost behind every test and the latter shows how to balance costs and tests writing because they are sensible aspects.

That being said, the most challenging thing to do when writing tests is to list all the use cases that need to be covered. The checklist I follow involves writing down a minimal test list and defining the priorities, as follows:

1. Add basic tests to ensure the correct application loading.

2. Write at least one success case covering the *happy path* for every endpoint. The happy path is the more straightforward case when no errors happen.

3. Write at least one failure case covering the *unhappy path* for every endpoint. As you can imagine, the unhappy path covers the most common errors, such as a wrong password when a user tries to log in.

4. The business cases should support you in defining the most common scenarios, such as user registration and unsubscribing a workflow.

Now we know *why* we are going to write tests and *which* tests we need, but we must also understand *how* we can write them. Let's find out in the next section.

How to write tests?

Implementing a test case using "pure" Node.js is simple, and it helps us understand the basic concepts of testing.

Let's create an `example.js` file, where we test that a variable is equal to the number `42`:

```
const assert = require('assert')
const myVar = '42'
assert.strictEqual(myVar, 42)
```

Running this file by `node example.js` will produce an error output:

```
AssertionError [ERR_ASSERTION]: Expected values to be strictly equal:
'42' !== 42
```

Congratulations! You have written your first failing test!

This step is essential during the test implementation; *the tests must fail* in the first place. When you have "red" tests, you define the input and the expected output. By doing so, you can start writing your code and verify that the implementation fits the desired result.

> **Test-driven development**
>
> Writing failing tests is called the "red phase" in **Test-Driven Development** (**TDD**). This is a methodology created by Kent Beck to write better and well-tested software applications. Even if you don't know or don't use this process, you can benefit from it by applying some simple rules, such as the one we have seen in this section.

In our example, the `myVar` declaration corresponds to our implementation, and we spotted the first bug in the code – the `'42'` string is not equal to the number `42`! Fixing the code and rerunning the file will not produce any output; this means that the test is successful.

The takeaway here is that our test files are made of a long list of **assertions** that define our expectations. When these assertions are successful, we can say that the tests are passing or are green.

The other notion to know about testing is the **coverage**. This gives you insight into how much of your code is tested. Let's see an example of updating our previous test file:

```
const assert = require('assert')
const myVar = 42
if (typeof myVar === 'string') {
  assert.strictEqual(myVar, '42')
} else {
  assert.strictEqual(myVar, 42)
}
```

The previous snippet is a simple code example to help you become confident with the concept of coverage; the `if` statement will always be `false`, as we expect.

Now, we need to use an external tool to see the coverage in action. To be specific, we are going to use the `istanbul` coverage tool and its command-line interface called `nyc`:

```
npx nyc node example.js
```

The preceding command downloads the `nyc` module, thanks to `npx`, and runs `node example.js`. As a result, we get a nice clear report:

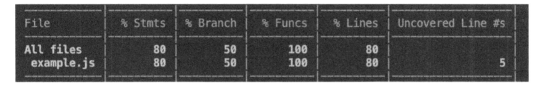

Figure 9.1 – The coverage report

This report tells us a lot about our test execution:

- `% Stmts`: The percentage of the statement executed during the run.
- `% Branch`: The percentage of tested paths the code can follow. A branch is created using conditional statements (`if`, the ternary operator, and `&&` and `||` operators), conditional loops (such as `while`, `for`, and `do-while`), `break` and `continue` statements, or a function's default parameters.
- `Uncovered Line #s`: The source code lines have not been executed.

To learn more about the output report, you can visit the `https://istanbul.js.org/` site.

> **100% coverage does not mean bug-free**
> Reaching 100% code coverage can be a complex task and sometimes a waste of time. You don't need to struggle to build up the coverage to consider your application tested. You should focus on the test cases that matter to your application; one more functional test is better than a not-so-useful test to reach 100% coverage.

The coverage report is handy to analyze; you can spot dead code branches that need to be removed or use cases that you forgot to list and check. This way, the code base maintenance is simplified, and it is possible to monitor the report results, while the test suite becomes larger.

Now you have learned the basic concepts about testing – assertions and coverage. All the Node.js testing frameworks provide these two capabilities. When you compare the packages to choose the best for you, it is highly suggested that you start evaluating the assertions that the framework implements and the coverage reports it generates.

We are now ready to write our application's tests in the next section.

Testing the Fastify application

We are ready to implement the tests for our Fastify application. Before we start, we need to choose a framework that will help us write the code. Let's complete this task first!

Installing the test framework

There are a lot of testing frameworks in the Node.js panorama. Some of them are highly opinionated, and others try to be agnostic. We are not going to discuss comparing the most famous modules. It is worth mentioning the ones most used by the community (in alphabetical order):

- AVA: `https://www.npmjs.com/package/ava`

- Jest: `https://www.npmjs.com/package/jest`

- Mocha: `https://www.npmjs.com/package/mocha`

- node-tap: `https://www.npmjs.com/package/tap`

- Tape: `https://www.npmjs.com/package/tape`

The framework we will use is `node-tap` because it has all the key features out of the box without needing extra configuration, such as the following:

- An easy-to-use and solid implementation

- Comprehensive assertions

- Parallel test execution

- An HTML coverage report format

It is simple to use, and the source code is easy to read, so it is perfect.

> **Choose wisely**
>
> You may be tempted to use the most downloaded testing framework from the previous list, such as Jest or Mocha. However, you must be aware that those frameworks were not designed to test server-side Node.js code. They heavily rely on implicit globals – an antipattern for testing predictable software output. This may impact your developer experience: `https://github.com/fastify/help/issues/555`.

Nonetheless, `node-tap` is largely used by the Fastify team. This might seem piddling, but it's worth remembering – sometimes, the plugins' documentation doesn't show a complete code example, but there are certainly some tests you can read to get more information and a full working example.

We need to install the `node-tap` dependency by running `npm install tap@15 --save-dev`. Let's see how to use its API.

Creating your node-tap cheatsheet test

Before writing the application's tests, we need to learn how to use `node-tap`. So, let's create a comprehensive test file to become familiar with all the most used features.

Create a new `test/cheatsheet.test.js` file and write this code:

```
const t = require('tap')
t.test('a test description', t => {
  t.plan(1)
  const myVar = '42'
  t.equal(myVar, 42, 'this number is 42')
})
```

The previous code shows a minimal test case that already teaches us a couple of good practices. After importing the module, we can start defining our test cases using the `t.test()` method. We must provide a test description and a test function that implements the logic to verify that our code is working.

The test function accepts one single argument. It is a `node-tap` object that provides you with the assertion methods to implement the test logic. `t.plan()` is a mandatory setup; you must declare how many assertions your test case will execute. If you don't set it, the run will fail.

> **Skip the test's plan**
>
> The `t.plan()` method is the best approach to ensure that all the assertions have been checked. It is instrumental when you are testing asynchronous code in callback style. Whenever you don't know how many assertions your code will execute exactly, for your convenience you can use `t.end()` instead of setting `t.plan()` in the initial implementation phase.

Running the script is as easy as running `node test/cheatsheet.test.js`. You will see some nice output that shows all the successful and failed steps. The code example should fail though. Try to fix it and rerun the file. Spoiler: `myVar` is not a number, as we learned in the previous *How to write tests* section.

The smart reader should have spotted that the `t.equal()` assertion function follows a strict comparison. Almost all the `node-tap` assertion functions accept three parameters in the following order:

1. What value do you want to compare?
2. What value do you expect?
3. An optional string message or option object.

Let's check the most used assertion functions by adding new test cases in our test file.

To check that a JSON object is like another, we can't use t.equal, which is used to compare values. We must use the following method instead to compare every object's field:

```
const sameStructure = { hello: 'world' }
t.strictSame(sameStructure, { hello: 'world' }, 'the
object is correct')
```

We want to check that a JSON object has some fields in some other cases. This is why we must use the match function:

```
const almostLike = {
  hello: 'world',
  foo: 'bar'
}
t.match(almostLike, { hello: 'world' }, 'the object is
similar')
```

We have checked that the almostLike variable has at least the hello property field. The t.match() assertion is more powerful than the previous example. It processes every regular expression against all the JSON input's enumerable fields. You can try it by setting a RegExp Node.js object:

```
t.match(almostLike, {
  hello: 'world',
  foo: /BAR/i
}, 'the object is similar with regex')
```

In the past example, the foo input property must match the /BAR/i regular expression.

For the last assertion functions, we will see the boolean checks and the nested tests:

```
t.test('sub test', function testFunction (t) {
  t.plan(2)
  const falsyValue = null
  t.notOk(falsyValue, 'it is a falsy value')
  t.test('boolean asserions', subTapTest => {
    subTapTest.plan(1)
    subTapTest.ok(true, 'true is ok')
  })
})
```

In this code snippet, you can read the t.notOk() and subTapTest.ok() functions that pass whether the value is falsy or truthy respectively. Moreover, the example shows a t.test() call inside another t.test(). This sub-test enables you to organize your use cases better and group them into logical steps. Note that a t.test() sub-test counts as an assertion when the plan counter is set up.

We must talk about one more thing before proceeding with the next section. So far, we have seen synchronous functions. What about `async` functions instead? `node-tap` makes it easy:

```
const fs = require('fs').promises
t.test('async test', async t => {
  const fileContent = await fs.readFile('./package.json',
    'utf8')
  t.type(fileContent, 'string', 'the file content is a
    string')

  await t.test('check main file', async subTest => {
    const json = JSON.parse(fileContent)
    subTest.match(json, { version: '1.0.0' })
    const appIndex = await fs.readFile(json.main, 'utf8')
    subTest.match(appIndex, 'use strict', 'the main file is
    correct')
  })
  t.pass('test completed')
})
```

Looking at the code example, we have provided an `async` function as a parameter. We do not need to set `t.plan` or call the `t.end` function. By default, the test ends when `Promise` returned by the `async` operation is fulfilled successfully. If `Promise` is rejected, the test fails.

Another note about the code is that the `t.test()` function returns `Promise` that can be awaited. This can be helpful to run some tests serially. We will delve more into this in the *Speeding up the test suite* section.

You can rely on the new `t.pass()` assertion in order to be sure about the code execution order, because it serves as a milestone that our code must meet.

We have learned how to use `node-tap` and write the assertions we will use to test the Fastify application. We did not cover all the assertions at our disposal. For further information, you can take a look at the official documentation: `https://node-tap.org/docs/api/asserts/`.

Now, let's move on to the next section and practice what we have learned so far.

How to write tests with Fastify?

We are ready to write out the first application test! Writing good tests is a process and it requires starting from a basis to build a solid structure. We will iterate multiple times on the same file to improve the test suite repeatedly.

As mentioned in the *Starting an optimal project* section of *Chapter 6*, we need to think about the tests we are going to write first. The most basic questions we could ask with our test are as follows:

- Does the application start correctly?

- Are the routes ready to listen to incoming HTTP requests?

- Does the application manage the unhealthy system?

To answer these questions, we can write the `test/basic.test.js` file:

```
const t = require('tap')
t.todo('the application should start', async (t) => {})
t.todo('the alive route is online', async (t) => {})
t.todo('the application should not start', async (t) => {})
```

We have listed the test cases we will write using the `t.todo()` method instead of the usual `t.test()`. The "todo" acts as a reminder to find those functions that need to be implemented.

Let's start with the first test. We need to load the application, but we are using the `fastify-cli` module to load our `app.js` file. Luckily, there is the `fastify-cli/helper` utility that helps us:

```
const fcli = require('fastify-cli/helper')
const startArgs = '-l info --options app.js'
t.test('the application should start', async (t) => {
  const envParam = {
    NODE_ENV: 'test',
    MONGO_URL: 'mongodb://localhost:27017/test'
  }
  const app = await fcli.build(startArgs, {
    configData: envParam
  })
  t.teardown(() => { app.close() })
  await app.ready()
  t.pass('the application is ready')
})
```

First of all, we need to load the `fastify-cli/helper` utility. It works as the `fastify` command in the `package.json > scripts` tag. In fact, we need the `startArgs` constant that includes the same configuration we use to start the server. You can copy and paste it from the `start` script in `package.json`.

The `fcli.build()` method is an asynchronous function that returns the Fastify instance. It is the same as running `fastify start <args>` from the shell. The key difference is that the server is not listening. We have already seen the difference between `fastify.ready()` and `fastify.listen()` in *Chapter 2*.

Note that the `build` function accepts an extra JSON parameter. The second option argument accepts a `configData` parameter, `opts`, received as input in the `app.js` file:

```
module.exports = async function (fastify, opts) { … }
```

This technique is one of the best practices to inject all the possible configurations the application supports. We will test every possible combination without building odd algorithms to load files using pattern matching.

To complete the test setup, we need to edit the `plugins/config.js` file by adding the following line:

```
await fastify.register(fastifyEnv, {
  confKey: 'secrets',
  data: opts.configData,
  schema: fastify.getSchema('schema:dotenv')
})
```

By doing so, we can control the `@fastify/env` data source. By default, the plugin reads data from `process.env`. By adding the `data` parameter to the plugin, it will be preferred over `env`, so it is possible to run the tests controlling the environment variables.

After building the Fastify instance, we need to track that we want to close the server when the test ends. The `t.teardown()` method accepts a function executed when this condition is met. If we forget to add it, the test script will never end because the open connections to the database will keep the Node.js runtime up and running.

Looking at the last two lines of the test implementation, we run `await app.ready()` to load the application without starting the HTTP server. All the plugins and routes are loaded only if the load itself is completed successfully. This test helps us find system-wide errors, such as broken plugins and misconfiguration.

Before starting the tests, we must not forget to turn on MongoDB by running `npm run mongo:start`. Now, the `node test/basic.test.js` command leads to a successful test run.

You have completed the first important step in a battle-tested Fastify application! Let's now implement the other tests case, focusing on improving our code step by step.

The second test case involves checking that the base route is up and running. The first challenge is to avoid repeating code to build the Fastify instance. It would be easy to copy and paste, but look at the following code:

```
async function buildApp (t, env = envParam, serverOptions){
  const app = await fcli.build(startArgs,
    { configData: env },
    serverOptions
  )
  t.teardown(() => { app.close() })
```

```
    return app
  }
  t.test('the alive route is online', async (t) => {
    const app = await buildApp(t)
    const response = await app.inject({
      method: 'GET',
      url: '/'
    })
    t.same(response.json(), { root: true })
  })
```

We have defined a `buildApp` function that reduces the complexity of building the application. It requires the test object as input and an optional JSON object that simulates the environment variables. Moreover, it is possible to merge the `configs/server-options.js` file's content with the `serverOptions` object, by providing a third parameter for the `fcli.build` function. It is really handy to be able to control every aspect of your Fastify server when running the tests.

The new notable utility is `app.inject(options[, callback])`. This function is a Fastify instance method that starts the server with a ready status, and it creates a fake HTTP request on the Fastify server. This means that the application is fully loaded, but the HTTP server is not listening for incoming HTTP requests. Nonetheless, you can call your routes, injecting a fake HTTP request. The fake HTTP request simulates a real HTTP one, creating `http.IncomingMessage` and `http.ServerResponse` objects.

Having an application up and running without holding the host's port has great advantages:

- The tests run faster
- It is possible to run tests simultaneously

The `app.inject()` interface accepts a JSON argument to compose the HTTP request. The most used properties are as follows:

- `method`: The request's HTTP method.
- `url` or `path`: The URL to call during the fake request.
- `headers`: A key-string JSON that sets the request's headers.
- `payload` or `body`: This can be a string, buffer, stream, or JSON object. The latter will be stringified and the content-type header will be set as `application/json` by default.
- `query`: A key-string JSON to set the request's query string.
- `cookies`: A key-string JSON to attach to the request cookies' headers.

You can find a complete options list by reading the `light-my-request` documentation at `https://github.com/fastify/light-my-request#injectdispatchfunc-options-callback`.

The `inject` API returns `Promise` when a callback is missing – as the `app.ready()` does. `Promise` fulfills an enhanced `http.ServerResponse` object that you can read in a simplified way – for instance, we used `response.json()` to get a JSON in our code example. This works only if the application returns a JSON payload. Other than that, you can access these properties:

- `statusCode`: Returns the HTTP status code response number
- `headers`: Returns a key-string JSON object that maps the response's headers
- `payload`: Returns the response's payload as a UTF-8 string
- `rawPayload`: Returns the response's payload as a `Buffer` object
- `cookies`: Returns a key-string JSON, mapping the response's cookie headers

The `light-my-request` module makes the application testing friendly and easy, thanks to the listed option utilities reducing the code we must write and incrementing what we can assert.

The previous code snippet that implements the route test can be summarized in these three steps:

1. Building the application.
2. Making a fake HTTP call.
3. Checking the response output.

I hope you are as excited about it as I am! In a few code lines, we have been able to spin up the whole application and make HTTP calls to the Fastify server.

Last but not least, we must deal with error cases. Testing the application's happy path could become boring at some point in the future. The real challenge is to verify how the software deals with errors, such as disconnection or wrong data input. We will complete our basic tests by adding some sub-tests to the `t.todo('the application should not start')` case.

We will force a bad environment to check that the server does not start when the configuration is wrong, but we must also be sure that the error tells us the correct information. Let's jump to the following code:

```
t.test('the application should not start', async mainTest => {
  mainTest.test('if there are missing ENV vars', async t => {
    try {
      await buildApp(t, {
        NODE_ENV: 'test',
        MONGO_URL: undefined
      })
      t.fail('the server must not start')
```

```
    } catch (error) {
      t.ok(error, 'error must be set')
      t.match(error.message, "required property
      'MONGO_URL'")
    }
  })
})
```

In this code snippet, we have added a sub-test that verifies that the `@fastify/env` plugin reacts correctly when the preconditions are not satisfied. The `t.fail()` method works as a trap; when executed, it stops the test and sets it as failed. By doing so, you are declaring that "this code must not execute."

Another useful test to add is an unreachable `mongodb` connection:

```
mainTest.test('when mongodb is unreachable', async t => {
  try {
    await buildApp(t, {
      NODE_ENV: 'test',
      MONGO_URL: 'mongodb://localhost:27099/test'
    })
    t.fail('the server must not start')
  } catch (error) {
    t.ok(error, 'error must be set')
    t.match(error.message, 'connect ECONNREFUSED')
  }
})
```

If you set the `MONGO_URL` incorrectly, the Fastify server will not be able to connect and not start as expected.

Well done! You have now written your first basic application tests. There is much more to consider in order to declare your application tested, but now you have a solid knowledge of the tools and methodology to add more and more checks.

In the next section, we will improve code reusability and the developer experience to write tests as smoothly as possible.

How to improve the developer experience?

Before writing all the test cases for the application's endpoints, we must ask ourselves whether there are repetitive tasks we must automate. In fact, we need to remember to start the MongoDB container and stop it every time. For this reason, we can optimize this process by adding a simple script to run before our tests. We will use the `dockerode` module (`https://www.npmjs.com/package/dockerode`) for this purpose.

Let's create a file named `test/helper-docker.js` and then map the `docker run` command into a configuration:

```
const Containers = {
  mongo: {
    name: 'fastify-mongo',
    Image: 'mongo:5',
    Tty: false,
    HostConfig: {
      PortBindings: {
        '27017/tcp': [{ HostIp: '0.0.0.0', HostPort:
        '27017' }]
      },
      AutoRemove: true
    }
  }
}
```

The previous code will be useful to control all the Docker images we may need in the future. The code snippet replicates the `docker run -d -p 27017:27017 --rm --name fastify-mongo mongo:5` command as a Node.js script that is easier to maintain and read.

Now, the software should be able to do the following:

- Understand whether MongoDB is running
- Start the MongoDB container
- Stop the MongoDB container

At this point, we can define the interface exported by the `helper-docker.js` file:

```
const Docker = require('dockerode')
function dockerConsole () {
  const docker = new Docker()

  return {
    async getRunningContainer (container) {
      // TODO
    },
    async startContainer (container) {
      // TODO
    },
    async stopContainer (container) {
      // TODO
    }
  }
```

```
  }
module.exports = dockerConsole
module.exports.Containers = Containers
```

We can use the `dockerode` module to implement the utility function to know it. To get the running container, the script must read all the running containers and then look for the one we are interested in:

```
async getRunningContainer (container) {
  const containers = await docker.listContainers()
  return containers.find(running => {
    return running.Names.some(name =>
    name.includes(container.name))
  })
},
```

The `container` argument will be `Containers.mongo`, defined in the code snippet at the beginning of this section.

To implement the function that starts the container instead, we need to pass the same `Containers.mongo` object to `dockerode`:

```
async startContainer (container) {
  const run = await this.getRunningContainer(container)
  if (!run) {
    await pullImage(container)
    const containerObj =
    await docker.createContainer(container)
    await containerObj.start()
  }
},
```

The `startContainer` function will start the MongoDB server locally if it is not already running – nice and easy!

Finally, the last `stopContainer` function utility to stop the container will look as follows:

```
async stopContainer (container) {
  const run = await this.getRunningContainer(container)
  if (run) {
    const containerObj =
    await docker.getContainer(run.Id)
    await containerObj.stop()
  }
}
```

We have completed the `docker` utility, but we still need to use it in our tests. To do it, we need to update the `basic.test.js` source code by adding the following script:

```
const dockerHelper = require('./helper-docker')
const docker = dockerHelper()
const { Containers } = dockerHelper
t.before(async function before () {
  await docker.startContainer(mongo)
})
t.teardown(async () => {
  await docker.stopContainer(dockerHelper.Containers.mongo)
})
```

You should be able to recognize three steps in the script:

1. Load the `helper-docker` script and initialize the variables.
2. Set the `t.before()` function that will start the MongoDB container. The `before` function will run once and before all the `t.test()` functions.
3. The `teardown` function will stop the container when all the tests are completed.

Now, whenever you and your team need to run the tests, it will no longer be necessary to remember anything more than just executing `npm test`.

The application's tests have not been completed yet, nor has the source code refactoring. This process is a continuous evolution and still requires some iteration before becoming stable. We have written just one single test file. The next challenge will be to write a new test file, without duplicating the source code. So, let's complete our tests in the next section.

Dealing with complex tests

So far, we have seen simple test cases that did not require multiple API requests. So, let's create the `test/login.test.js` file to verify the sign-up and first user login to our application. We will use what we have learned so far, keeping in mind that we don't want to replicate code.

We need to build the Fastify instance to write new test cases, as we did in the `test/basic.test.js` file. To do so, we need to do the following:

1. Create a new utility file and call it `test/helper.js`.
2. In the `test/helper.js` file, move the `buildApp` function and its configuration variables, `startArgs` and `envParam`. This action requires some copying and pasting.
3. Update the `test/basic.test.js` file within the new import, `const { buildApp } = require('./helper')`.

By doing so, we can reuse code to instantiate the Fastify application across all the test files we are going to create. We are now ready to write more complex tests.

Reusing multiple requests

Every route has data requirements that can be satisfied by replicating a client's workflow – for example, we need to create a user first to test user deletion. So, we can start writing test cases for the login process. The test suite should answer these questions:

- Does the authorization check block unauthorized users?

- Do the register and login endpoints work as expected?

The first check should verify that the protected routes are protected, so we can implement a bit more logic into the test's source code:

```
t.test('cannot access protected routes', async (t) => {
  const app = await buildApp(t)
  const privateRoutes = [ '/me' ]
  for (const url of privateRoutes) {
    const response = await app.inject({ method: 'GET', url
    })
    t.equal(response.statusCode, 401)
    t.same(response.json(), {
      statusCode: 401,
      error: 'Unauthorized',
      message: 'No Authorization was found in
      request.headers'
    })
  }
})
```

We can iterate the `privateRoutes` array to check that the routes are secured. Here, I'm showing how to automate code without repeating yourself.

Before logging into the application, the user must register to the platform, so we should add a test for it. This is a simple task nowadays, but here is the code for completeness:

```
t.test('register the user', async (t) => {
  const app = await buildApp(t)
  const response = await app.inject({
    method: 'POST',
    url: '/register',
    payload: {
      username: 'test',
      password: 'icanpass'
```

```
    }
  })
  t.equal(response.statusCode, 201)
  t.same(response.json(), { registered: true })
})
```

Then, we need to test the login endpoint to verify that it works as expected. It must return a **JWT token** (as discussed in *Chapter 8*), and we can use it to access the `privateRoutes` endpoints. As you can imagine, the login test is straightforward, as follows:

```
t.test('successful login', async (t) => {
  const app = await buildApp(t)
  const login = await app.inject({
    method: 'POST',
    url: '/authenticate',
    payload: {
      username: 'test',
      password: 'icanpass'
    }
  })
  t.equal(login.statusCode, 200)
  t.match(login.json(), { token: /(\w*\.){2}.*/ })
})
```

The authentication test executes a POST call, providing the correct user's data and verifying that the service returns a token string. You can apply a stricter validation to the output token as well.

Now, we can use the generated token by adding a new sub-test after the `t.match()` assertion:

```
t.test('access protected route', async (t) => {
  const response = await app.inject({
    method: 'GET',
    url: '/me',
    headers: {
      authorization: `Bearer ${login.json().token}`
    }
  })
  t.equal(response.statusCode, 200)
  t.match(response.json(), { username: 'John Doe' })
})
```

The `access protected route` test relies on the `login` object to authenticate the request and successfully access the endpoint. Note that the sub-test does not need to build the application; we can use the one created by the parent test case.

It is possible to create complex workflows and simulate every scenario to cover the business cases.

Mocking the data

To complete our test journey, we must talk about **mocks**. A mock is a fake implementation of a real application's component that acts conditionally to simulate behaviors that would be hard to replicate. We will use a mock to test a failed registration while the service inserts data into a database.

Many tools help you write mocks, but we will keep it at a low level to understand how they work. Let's jump into the code:

```
function cleanCache () {
  Object.keys(require.cache).forEach(function (key) {
  delete require.cache[key] })
}
t.test('register error', async (t) => {
  const path = '../routes/data-store.js'
  cleanCache() // [1]
  require(path) // [2]
  require.cache[require.resolve(path)].exports = { // [3]
    async store () {
      throw new Error('Fail to store')
    }
  }
  t.teardown(cleanCache) // [4]
  const app = await buildApp(t)
  const response = await app.inject({ url: '/register', … })
  t.equal(response.statusCode, 500)
})
```

Mocks rely on how Node.js loads the source code of an application. For this reason, we need to take over the default logic for our scope. Every time a `require('something')` statement runs, a global cache is fulfilled within the `module.exports` exported data, so if you run the `require` statement twice, the file is loaded just once. That being said, we need a function to clean this cache to inject our mock implementation. We need the `cleanCache` function that removes all the loaded code. This is not performant at all. You could filter the output based on your project path to optimize it.

The test implementation does a few things before calling the `buildApp` function (as seen in the preceding code block):

1. [1] cleans the cache; we need to remove all the cached files that use the `path` file. It is unmanageable to know every file that uses it, so we will clean the whole cache as a demonstration.

2. At [2] we load the target file we are going to mock.

3. The [3] applies the mock to the cache. To do it, we need to know the `path` file interface. As you can understand, this is an invasive test that does not adapt to any refactor.

4. Finally the [4] block must remove the mock implementation when the test ends to let Node.js reload the original file. We can't clean just the `path` cache because all the files that used the mock have been cached within the mock itself.

As you can see, the mock testing technique requires knowledge of the Node.js internals. The modules that help you mock your code work as the previous code snippet but provide a better user experience. Moreover, this test may change over time when the `path` file changes. You should evaluate whether you need to couple your tests within the code base.

Sometimes, this technique is not an option. An example is a third-party module that you don't need to test in your test suite, such as an external authorization library.

Now you have added new tools to your test toolkit that you will use to evaluate more options during your test suite implementation. We have slowed down the test by clearing the cache in the previous section. Let's find out how to speed up the tests in the next section.

Speeding up the test suite

There are not many tests in the actual application, but they're all pretty fast. While your project will grow, the tests will become more and more time-consuming and more annoying to run. It is not uncommon to have a test suite that runs in a span of 15 minutes, but that is too much time! Now, we are going to see how to speed up a test run to avoid a situation like this, by parallelizing the tests' executions and evaluating the pitfall this technique carries.

Running tests in parallel

To improve our test run, we need to update the test script in `package.json`:

```
"test": "tap test/**/**.test.js",
```

The `npm test` command will execute all the files in the `test/` folder that ends with the `test.js` suffix. The cool thing is that each file runs in parallel on a dedicated Node.js process! That being said, it hides some considerations you must be aware of when writing tests:

*   `process.env` is different for every test file
*   There are no shared global variables across files (and tests)
*   The `require` module is performed at least once per Node.js process spawned
*   Executing `process.exit()` will stop one execution file

These are not limitations, but having these rules helps you to organize code in the best way possible, and to run tests the fastest. Moreover, you are forced to avoid global objects and functions that add side effects (`https://softwareengineering.stackexchange.com/questions/40297/what-is-a-side-effect`). For this reason, the factory pattern we have adopted since the first chapter is a big win – every test case will build its own objects with its own configuration, without conflicting with other files.

> **The --jobs argument**
>
> The `tap` command interface accepts a `-j=<n>` `--job=<n>` parameter that sets how many test files can be run in parallel. By default, it is set as the system's CPU core count. Setting it to `1` disables the parallelism.

The `node-tap` framework has a comprehensive section about running tests in parallel at `https://node-tap.org/docs/api/parallel-tests/`.

How to manage shared resources?

Managing shared resources is a parallelism con. We need to implement the last refactor on our test suite to achieve this result. The shared resource I'm talking about is a database. Using the `helper-docker` utility in every test file is not an option. We would face errors due to the host's port already in use or a Docker conflict, such as the following:

```
Conflict. The container name "/fastify-mongo" is already in use by
container "e24326". You have to remove (or rename) that container to
be able to reuse that name
```

There are some options to solve this issue:

- Customize the configuration of each test file. Running a database container for every file requires a lot of system resources, so you must evaluate this option carefully. It is the easiest way to fix the issue, but we would slow down the suite as a result.

- Change the database. Right now, we are turning on an actual database, but there are many alternatives in the npm ecosystem, such as in-memory databases that emulated NoSQL databases or SQL ones. This is for sure a good option you must take into account.

- Create the pre-test and the post-test scripts to spin up the shared resources before the tests' execution. Note that every file needs its own dataset or database schema to border the assertions, or one test could erase all the data for other tests!

These are the most common solutions to the shared resources problem. The first option does not work with limited resources. The second option does not work if you use a database that does not have an in-memory implementation. So, we will implement the third option because it teaches you one real scenario that works every time. Don't be afraid. It is a matter of refactoring the source code a bit.

Let's create a new `test/run-before.js` file; cut and paste the `before/teardown` code from the `test/basic.test.js` file. The new file outcome will be as follows:

```
const t = require('tap')
const dockerHelper = require('./helper-docker')
const docker = dockerHelper()
const { Containers } = dockerHelper
t.before(async function before () {
  await docker.startContainer(Containers.mongo)
})
```

The `basic.test.js` file will be smaller and smaller at every refactor. It means we are doing great. Now, we need another file called `test/run-after.js`. It will be similar to the `run-before` one, but in place of `t.before()`, we must cut the `teardown` function:

```
t.teardown(async () => {
  await docker.stopContainer(Containers.mongo)
})
```

We are almost done with our refactoring. Now, we must update the `basic.test.js` file by updating all the `buildApp` usages and setting the default database:

```
const app = await buildApp(t, {
  MONGO_URL: 'mongodb://localhost:27017/basis-test-db'
})
```

Then, it is the `login.test.js` file's turn to set up its own database instance:

```
const app = await buildApp(t, {
  MONGO_URL: 'mongodb://localhost:27017/login-test-db'
})
```

Finally, we need to use two new `node-tap` arguments by editing `package.json`:

```
"test": "tap --before=test/run-before.js test/**/**.test.js
--after=test/run-after.js",
"test:nostop": "tap --before=test/before.js test/**/**.test.js",
```

The `--before` parameter will execute the input file before the whole test suite. The `--after` argument does the same but at the end of the test suite run. Note that the `test:nostop` addition is equal to the `test` script but does not stop and clean the database server at the end of the process. This script is really helpful when you are developing and you need to check the data on your database manually.

Do you find it difficult to manage shared resources? If yes, then thanks to the Fastify coding style pattern, you should become very comfortable with these refactors. We can only do so because there are no global objects, and we can instantiate as many Fastify instances as we need without caring about the host's ports.

Now, you have the initial knowledge to deal with parallelism complexity. It is not easy, but you can overcome the complexity with clear code and reusable functions.

In the next section, we will provide some suggestions to push your code base to the stars.

Where tests should run

Up until now, we have executed our test manually on our PC. That is fine, and it is mandatory during the development phase. However, this is not enough because our installation could be edited, or we could have some uncommitted files.

To solve this issue, it is possible to add a **Continuous Integration** (**CI**) pipeline that runs remotely to manage our repository. The CI pipeline's primary duties are as follows:

- Running the test suite to check the code in the remote Git repository
- Building a code base to create artifacts if necessary
- Releasing the artifacts by triggering a **Continuous Delivery** (**CD**) pipeline to deploy the software

The CI workflow will notify us about its status, and if it is in a red state, the application's tests are failing with the last commit. Running the test remotely will prevent false-positive issues due to our local environment setup.

We will build a simple CI workflow by adopting GitHub Actions. This is a free service for public repositories, with a free limited quota for private ones. We will not go into detail and just take a quick look at how easy it is to start using a CI pipeline.

To create a CI workflow, you need to create a new file named `.github/workflows/ci.yml`. The source must be as follows:

```
name: CI
on: [push, pull_request]
jobs:
  test-job:
    runs-on: ubuntu-latest
    steps:
      - name: Check out the source code
        uses: actions/checkout@v2
      - name: Install Node.js
        uses: actions/setup-node@v2
        with:
          node-version: 18
      - name: Install Dependencies
        run: npm install
      - name: Run Tests
        run: npm test
```

As you can see, the script maps every step you should follow to run the project:

1. Check out the source code.
2. Install the desired Node.js version.
3. Install the project.
4. Run the test script.

This step-by-step process is crucial in a CI configuration. If you want to try other vendors, such as CircleCI, Bitbucket Pipelines, or Travis CI, you will need to change the configuration file's syntax, but the logic will be unaltered.

Committing the previous code example will trigger the GitHub action automatically. You can see it by looking at the repository's **Actions** tab, as shown in the following screenshot:

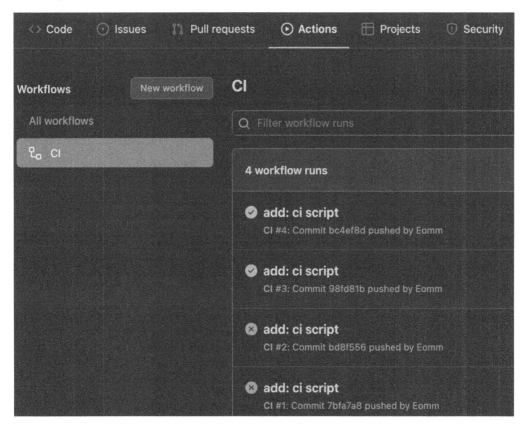

Figure 9.2 – The CI executions

As you can see in *Figure 9.2*, the workflow will fail on the first try. We need to fix our test script. So, we must read the console's output in order to understand what was not working properly.

The observant among you would have noticed this error output at the end of the npm test command, even if the tests were successful:

```
ERROR: Coverage for lines (85.41%) does not meet global threshold
(100%)
ERROR: Coverage for functions (91.66%) does not meet global threshold
(100%)
ERROR: Coverage for branches (0%) does not meet global threshold
(100%)
ERROR: Coverage for statements (85.41%) does not meet global threshold
(100%)
```

The error is due to a default node-tap configuration that requires 100% coverage. To reach this coverage level, we must add a new flag to the package.json's test script:

```
"test": "tap --before=test/run-before.js test/**/**.test.js
--after=test/run-after.js --no-check-coverage",
"test:coverage": "tap --coverage-report=html --before=test/run-before.
js test/**/**.test.js --after=test/run-after.js",
```

The --no-check-coverage argument solves the test failure due to its coverage below the 100% threshold.

The last addition to complete this journey into the node-tap framework and application tests is the test:coverage script, added in the previous code snippet. Running the script by executing the npm run test:coverage command should open your system's browser at the end, showing a nice HTML report as follows:

All files

85.41% Statements 41/48   **0%** Branches 0/2   **91.66%** Functions 11/12   **85.41%** Lines 41/48

Press *n* or *j* to go to the next uncovered block, *b*, *p* or *k* for the previous block.

Filter:

File ▲		Statements		Branches
fastify-todo-list-api		100%	7/7	100%
fastify-todo-list-api/configs		100%	1/1	100%
fastify-todo-list-api/plugins		78.12%	25/32	0%
fastify-todo-list-api/routes		100%	4/4	100%
fastify-todo-list-api/schemas		100%	4/4	100%

Figure 9.3 – A coverage HTML report

If the browser doesn't open automatically the web page, it is possible to open it manually by clicking on the `coverage/lcov-report/index.html` file, that has been generated in the project's root path during the test execution.

Figure 9.3 shows how you can build a coverage report that you can navigate using your browser. By clicking on the blue highlighted links, you will see every repository's file and how many times a single code line has been executed during the test execution:

All files / fastify-todo-list-api/routes routes.js

100% Statements 3/3   **100%** Branches 0/0   **100%** Functions 2/2   **100%** Lines 3/3

Press *n* or *j* to go to the next uncovered block, *b*, *p* or *k* for the previous block.

```
1        'use strict'
2
3   2x   module.exports = async function root (fastify, opts) {
4   3x     fastify.get('/', async function welcomeHandler (request, reply) {
5   2x       return { root: true }
6        })
7      }
8
```

Figure 9.4 – Source code coverage

The coverage output helps you understand what is not tested in your application, allowing you to make appropriate decisions.

Summary

This chapter is dense with information about new processes and tools. Now, you should be comfortable designing a test suite for a Node.js backend application. You should be able to evaluate a testing framework that fits your needs and boosts your productivity.

You have learned how to use `node-tap`, from basic assertions to advanced parallel test execution. Moreover, you can test a Fastify application and take advantage of Fastify's `inject` feature. You don't have to worry about testing your API's routes, whatever the level of complexity is.

Finally, we have seen how to integrate a CI pipeline using GitHub Actions and its logic to keep our repository away from regressions and production issues.

Now, you are ready to proceed to the next step and build a secure and reliable application. We mentioned CD earlier in this chapter; it is now time to see it in action in *Chapter 10*.

10
Deployment and Process Monitoring for a Healthy Application

We are building a new and shiny Fastify API and want to expose it on the internet to gather feedback before our official launch. When we look online, there are plenty of options… but what should we use? After that, we will then need to figure out how to monitor the health of our server (because we always monitor our applications, right?).

In this chapter, we will unpack the basics of a monolith deployment, using Docker, MongoDB, and Fly.io. We will also review the key Node.js metrics, how to extract them from our application with Prometheus, and then how to easily consult them on Grafana.

This is the learning path we will cover in this chapter:

- Testing our Docker image for deployment
- Hosting our DB on MongoDB Atlas
- Choosing a cloud provider
- Deploying to Fly.io
- Setting up continuous deployment
- Collecting application process data

Technical requirements

As mentioned in the previous chapters, you will need the following:

- A working Node.js 18 installation
- A text editor to try the example code
- Docker
- An HTTP client to test out code, such as cURL or Postman
- A GitHub account

All the snippets in this chapter are on GitHub at `https://github.com/PacktPublishing/Accelerating-Server-Side-Development-with-Fastify/tree/main/Chapter%2010`.

Testing our Docker image with a local deployment

In *Chapter 6*, we set up our application for usage with Docker. This is a critical step for our TODO List application. Now, it's time to test the image with Docker Compose to verify that everything is working as expected.

Let's recap our `DockerFile`, with a minor modification:

```
FROM node:18-alpine as builder
WORKDIR /build
COPY package.json ./
COPY package-lock.json ./
ARG NPM_TOKEN
ENV NPM_TOKEN $NPM_TOKEN
RUN npm ci --only=production --ignore-scripts
FROM node:18-alpine
RUN apk update && apk add --no-cache dumb-init
ENV HOME=/home/app
ENV APP_HOME=$HOME/node/
ENV NODE_ENV=production
WORKDIR $APP_HOME
COPY --chown=node:node . $APP_HOME
COPY --chown=node:node --from=builder /build $APP_HOME
USER node
EXPOSE 3000
ENTRYPOINT ["dumb-init"]
CMD ["./node_modules/.bin/fastify", "start", "-a", "0.0.0.0", "-l",
"info", "--options", "app.js"]
```

As you can see, we have modified CMD to directly call the `fastify` command with the following options:

- `-a 0.0.0.0` is fundamental to allowing the container to listen to all addresses
- `--options` makes sure we load the options from `app.js`
- `-l info` configures the logging level

Now, we can test this deployment using a local `docker-compose`, which is useful to verify that everything works well locally, as it is easier to debug than in the cloud. Specifically, we are going to connect our application to a MongoDB instance running within Docker as well.

Let's save the following as `docker-compose-test.yml`:

```yaml
version: "3.7"

services:
  mongo:
    image: mongo:5
    volumes:
      - data:/data/db

  app:
    build:
      context: ./
      dockerfile: Dockerfile
    ports:
      - "3042:3000"
    environment:
      JWT_SECRET: changethis
      MONGO_URL: mongodb://mongo:27017/todo

volumes:
  data:
```

Let's go line by line through the content of this file. First, we identify the version of the `docker-compose` file as different versions have different syntaxes. Second, we define two services: one is mongo, our database, and the other is app, which we build from the current folder. Note that in the app definition, we specify the environment variables our application needs:

- `JWT_SECRET`, which should be changed to secure your application.
- `MONGO_URL`, which identifies how we connect to our database. Note that we use mongo as a domain name – Docker will automatically resolve it to the host running our MongoDB!

Lastly, it's important to cover the concept of ports in Docker. In the `docker-compose-test.yml` file, we specify `"3042:3000"`: we map the TCP port `3042` of the host to port `3000` of the container. Therefore, we can now head to `http://localhost:3042/` to see our application running.

In this setup, persist the database in a data volume to avoid losing our data when the container is removed (such as when we update it).

Thanks to `docker-compose`, we can verify and test that our DockerFile works as expected. So, run the following:

```
docker-compose -f docker-compose-test.yml up
```

Docker will download the base images, build our image, and then execute our application (if you do not have `package-lock.json` in your folder, run `npm i` to generate it, otherwise, `docker build` will give an error). We can now `curl http://localhost:3042/` to verify that everything is working as expected.

> **MongoDB cluster**
>
> The MongoDB documentation (`https://www.mongodb.com/basics/clusters/mongodb-cluster-setup`) recommends using at least three MongoDB nodes because, in production, we want to have at least two replicas of the data to tolerate the failure of one node. In our "local" deployment, we are just using one – it's unsafe, but it's okay for our purpose.

Now, we want to move our setup to the cloud. To do that, first, we are going to use MongoDB Atlas to create a three-node MongoDB cluster.

Hosting our DB on MongoDB Atlas

As we write this, the simplest way to provision a MongoDB cluster is to use MongoDB Atlas, which will allow 512 MB of storage for free. To employ this, please follow the MongoDB Atlas setup tutorial – while we include the screenshot for this process here, the process might vary.

The first step is to sign up to MongoDB Atlas at `http://mongodb.com`. After you have completed the signup and email verification process, you can select the tier in which you want your new database to be created. We will select the **Shared** option, which is the free tier. After we have selected our tier, we now choose where we want our MongoDB instance to be located. Choose a location that's near to you!

Now, it's time to add the security mechanism for our database. Make sure you are in the **Quickstart** tab below **SECURITY**, as shown in *Figure 10.1*. For this book, select the **Username and Password** authentication method.

Security Quickstart

To access data stored in Atlas, you'll need to create users and set up network security controls. Learn more about securit

DEPLOYMENT

Database

Data Lake   PREVIEW

DATA SERVICES

Triggers

Data API

Data Federation

Atlas Search

SECURITY

Quickstart

Database Access

Network Access

Advanced

New On Atlas   1

How would you like to authenticate your connection?

Your first user will have permission to read and write any data in your project.

Username and Password	Certificate

Create a database user using a username and password. Users will be given the *read and write to any database privilege* by default. You can update these permissions and/or create additional users later. Ensure these credentials are different to your MongoDB Cloud username and password. You can manage existing users via the Database Access Page.

Username

[Enter username]

Password

[Enter password]   [🔍 Autogenerate Secure Password]   [📋 Copy]

[Create User]

Username	Authentication Type	
fastify	Password	[✏ EDIT]

Where would you like to connect from?

Enable access for any network(s) that need to read and write data to your cluster.

My Local Environment	Cloud Environment ADVANCED
Use this to add network IP addresses to the IP Access List. This can be modified at any time.	Use this to configure network access between Atlas and your cloud or on-premise environment. Specifically, set up IP Access Lists, Network Peering, and Private Endpoints.

Add entries to your IP Access List

Only an IP address you add to your Access List will be able to connect to your project's clusters. You can manage existing IP entries via the Network Access Page.

IP Address	Description		
[Enter IP Address]	[Enter description]	[Add Entry]	[Add My Current IP Address]

IP Access List	Description	
0.0.0.0/0	all internet	[🗑 REMOVE]

[Finish and Close]

Figure 10.1: How our database in MongoDB Atlas is configured

The most important configuration for this book is to enable connections from all IP addresses. This can be done in the **IP Address** field in the previous figure; enter 0.0.0.0/0, which will identify all IP addresses. While this is highly insecure, a strong password is more than enough for simple applications.

> **Other connection methods**
>
> Please refer to the MongoDB Atlas documentation for configuring more secure connection methods: https://www.mongodb.com/docs/atlas/security/config-db-auth/.

You'll now need to copy a connection string to your MongoDB database. This must be under an option similar to **Connect to your database** (we say "similar to" as we cannot say the exact option because it would probably become out of date at the time of reading).

Now, we can run the connection string in our terminal:

```
$ mongosh
"mongodb+srv://fastifytodolist.ypq0399.mongodb.net/myFirst
Database" --apiVersion 1 --username fastify
Enter password: ****************
Current Mongosh Log ID: XXXXXXXXXXX
Connecting
to:        mongodb+srv://<credentials>@fastifytodolist
.ypq0399.mongodb.net/myFirstDatabase?appName=mongosh+1.5.1
Using MongoDB:        5.0.9 (API Version 1)
Using Mongosh:        1.5.1

For mongosh info see: https://docs.mongodb.com/mongodb-shell/


To help improve our products, anonymous usage data is collected and
sent to MongoDB periodically (https://www.mongodb.com/legal/privacy-
policy).
You can opt-out by running the disableTelemetry() command.

Atlas atlas-lk17c9-shard-0 [primary] myFirstDatabase>
```

Everything worked as expected! We can now try to connect our application from our development machine to our new database in the cloud with the following:

```
MONGO_URL="mongodb+srv://USERNAME:PASSWORD@YOURDBDOMAIN/
myFirstDatabase" JWT_SECRET=supsersecret npm start
```

Alternatively, you can add the following and replace the value for MONGO_URL in `.env` file:

```
MONGO_URL="mongodb+srv://USERNAME:PASSWORD@YOURDBDOMAIN/
myFirstDatabase"
JWT_SECRET=supsersecret
```

Then in terminal, run `npm start`.

As we now have a cloud database, we can look to deploy our application to the cloud!

Choosing a cloud provider

Most people who start using a new technology wonder what cloud provider would provide the best experience to deploy their applications. In this book, we are not considering any solution that would be too much effort given the little space available.

Here is a list of providers that are worthwhile checking out:

- Heroku – it started supporting Node.js in 2011 and it's by far one of the most mature and stable products.

- Google Cloud Run – it's based on Knative (`https://knative.dev/`) and the Kubernetes stack.

- AWS Lambda – it's the original "serverless" runtime, enabling the executions of "functions" that can scale elastically. Every function only processes one request at a time: while this makes it easy to scale and operate, I/O heavy applications are at a disadvantage. AWS Lambda is based on Firecracker (`https://firecracker-microvm.github.io/`).

- Vercel – it's the deployment platform for frontend teams, based upon AWS Lambda.

- Fly.io – it's based on Firecracker, and as such, it allows for an extremely fast restart of processes and the ability to "scale to zero."

You can find a list of other serverless deployments at `https://www.fastify.io/docs/latest/Guides/Serverless/`.

Node.js and Fastify shine when used to serve multiple, parallel requests. This maximizes the idle moment in the event loop to process other requests. Therefore, the best choice is to use "full" servers that can be dynamically scaled based on the load.

> **Deploying on AWS Lambda**
>
> If you would like to deploy to AWS Lambda, it's possible to use `https://github.com/fastify/aws-lambda-fastify`. The advantage to native Lambda development is that you could develop your application as you normally would if it was deployed in a standard Node.js process.

For this book, we are going to cover Fly.io because it looks like the most innovative solution and provides easy automation for continuous deployment.

Deploying to Fly.io

Fly.io's main interface is a command-line tool called `flyctl`, which we can install with the following:

- `iwr https://fly.io/install.ps1 -useb | iex` on Windows PowerShell

- `curl -L https://fly.io/install.sh | sh` on Linux and macOS

- You can also use `brew install flyctl` on macOS too

Signing up with Fly.io is easy: issue the `flyctl auth signup` command. We recommend connecting your GitHub account, as you will need it later.

We can now deploy to Fly.io by executing `flyctl launch` in our current working directory (make sure there are no `fly.toml` files) and answering the following questions:

```
$ flyctl launch
Creating app in /path/to/Chapter 10
Scanning source code
Detected a Dockerfile app
? App Name (leave blank to use an auto-generated name):
? Select organization: Matteo Collina (personal)
? Select region: fra (Frankfurt, Germany)
Created app shy-fog-346 in organization personal
Wrote config file fly.toml
? Would you like to setup a Postgresql database now? No
? Would you like to deploy now? Yes
Deploying shy-fog-346
...
```

The first questions that `flyctl` asks are about our application: what account to use, where to start it, and whether we want a PostgreSQL database. Then, it builds our container and deploys our application to Fly.io.

Then, the script goes ahead and tries to deploy our application until it fails:

```
...
Preparing to run: `dumb-init ./node_modules/.bin/fastify start -l info
--options app.js` as node
2022/07/28 12:54:40 listening on [fdaa:0:582d:a7b:c07e:2d5f:9db0:2]:22
(DNS: [fdaa::3]:53)
Error: env must have required property 'MONGO_URL', env must have
required property 'JWT_SECRET'
    at loadAndValidateEnvironment (/home/app/node/node_modules/@
fastify/env/node_modules/env-schema/index.js:93:19)
```

```
    at loadAndValidateEnvironment (/home/app/node/node_modules/@
fastify/env/index.js:8:20)
    at Plugin.exec (/home/app/node/node_modules/avvio/plugin.
js:131:19)
    at Boot.loadPlugin (/home/app/node/node_modules/avvio/plugin.
js:273:10)
    at process.processTicksAndRejections (node:internal/process/task_
queues:82:21) {
  errors: [
    {
      instancePath: '',
      schemaPath: '#/required',
      keyword: 'required',
      params: [Object],
      message: "must have required property 'MONGO_URL'"
    },
    {
      instancePath: '',
      schemaPath: '#/required',
      keyword: 'required',
      params: [Object],
      message: "must have required property 'JWT_SECRET'"
    }
  ]
}2022-07-28T12:54:42.000 [info] Main child exited normally with code:
1
2022-07-28T12:54:42.000 [info] Starting clean up.
***v2 failed - Failed due to unhealthy allocations - no
    stable job version to auto revert to and deploying as v3

Troubleshooting guide at https://fly.io/docs/getting-started/
troubleshooting/
Error abort
```

As you can see in the error message, the server cannot find the JWT_SECRET and MONGO_URL environment variables. Let's add them as Fly.io secrets:

```
$ flyctl secrets set MONGO_URL="mongodb+srv://
YOURUSERNAME:YOURPASSWORD@YOURDATABASE.mongodb.net/myFirstDatabase"
...

$ fly secrets set JWT_SECRET="ALONGRANDOMSTRING"

Release v4 created
==> Monitoring deployment
...
./node_modules/.bin/fastify start -l info --options app.js` as node
```

```
[info]2022/07/28 13:19:02 listening on [fdaa:0:582d:a7b:66:b7
3e:6147:2]:22 (DNS: [fdaa::3]:53)
[info]
{"level":30,"time":1659014344254,"pid":517,"hostname":"b73e6147",
"msg":"Server listening at http://[::1]:3000"}
[info]
{"level":30,"time":1659014344262,"pid":517,"hostname":"b73e6147",
"msg":"Server listening at http://127.0.0.1:3000"}
--> v4 failed - Failed due to unhealthy allocations - no stable job
version to auto revert to and deploying as v5

--> Troubleshooting guide at https://fly.io/docs/getting-started/
troubleshooting/
```

Oh no! It's still failing. What happened? Let's open the `fly.toml` file (the Fly.io configuration file): there is an `internal_port = 8080` line that instructs Fly.io to route any incoming requests to port `8080` of our container. However, our application is being launched at port `3000`! Let's change the line to `internal_port = 3000` and issue `flyctl launch`.

Hurray! Our application is now deployed at `https://shy-fog-346.fly.dev/` (replace with your own URL).

In this section, we configured Fly.io and set two application secrets on Fly.io.

Setting up continuous deployment

We are setting up a continuous deployment system so that every commit pushed to the main branch of your GitHub repository is automatically deployed to Fly.io.

> **Automating your project with GitHub Actions**
>
> We recommend automating your development process with the use of GitHub Actions. They can be used to automatically run your automated tests, deploy your project, and even synchronize your issues with your other management software. Read more at `https://docs.github.com/en/actions`.

First, let's create a repository on GitHub, clone it locally, and push our code there! To upload our code to GitHub, we run `git clone`, `git add`, `git commit`, and `git push` in our terminal.

> **Using Git and GitHub**
>
> If you are not familiar with Git and GitHub, we recommend you to follow the GitImmersion tutorial at `https://gitimmersion.com/`, as well as `https://docs.github.com/en/get-started/quickstart/create-a-repo`, to create the repository on GitHub, clone it, and push our application.

To deploy to Fly.io automatically, we must install and configure the `flyctl` client. We will need to authenticate our GitHub Action to Fly.io and to do this, we need a Fly.io authorization token. This can be generated by running `fly auth token` in our terminal.

After that, copy that value and open our GitHub repository. Inside **Settings** | **Secrets** | **Actions**, open **New Repository Secret**. Then, set `FLY_API_TOKEN` as the name and our token as the value, and click on **Add secret**.

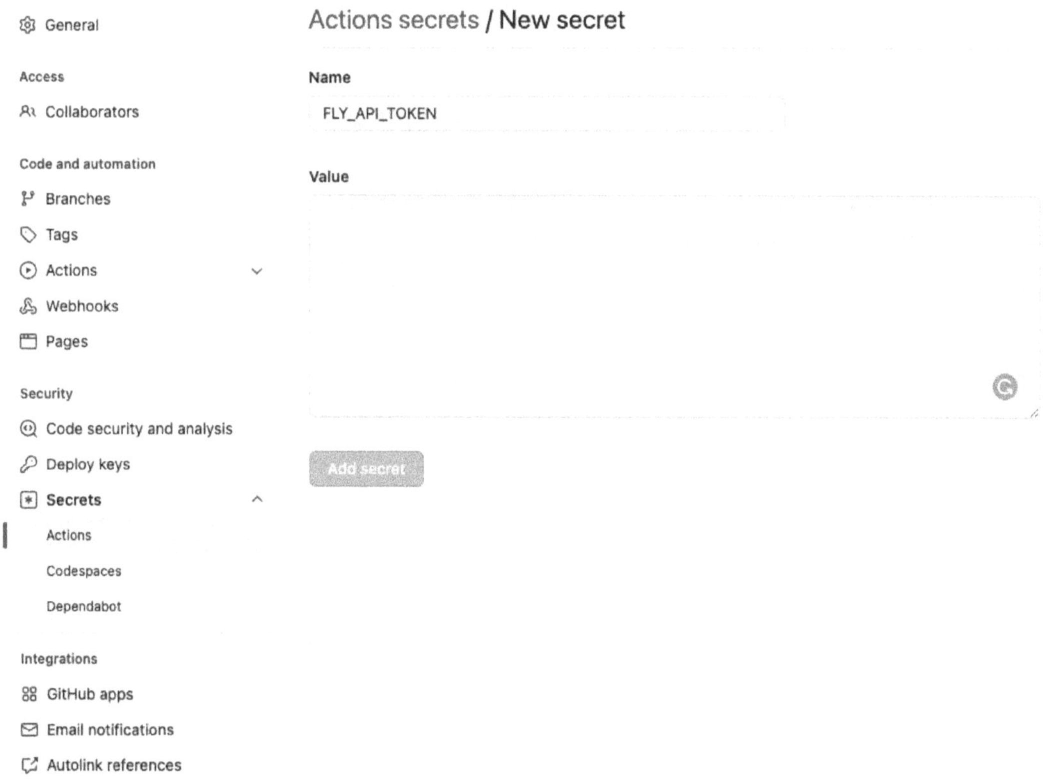

Figure 10.2: Adding the FLY_API_TOKEN as a GitHub Action secret

After configuring our secret, it's time to use GitHub Actions to automatically deploy our application. To do so, inside our working directory, create a `.github/workflows/fly.yml` file with the following content:

```
name: Fly Deploy
on:
  push:
    branches:
      - main
```

```
env:
  FLY_API_TOKEN: ${{ secrets.FLY_API_TOKEN }}

jobs:
  deploy:
    name: Deploy app
    runs-on: ubuntu-latest
    steps:
      - uses: actions/checkout@v3
      - uses: superfly/flyctl-actions/setup-flyctl@master
      - run: flyctl deploy –remote-only
```

This GitHub Action uses the `setup-flyctl` action to configure the Fly.io client and then uses `flyctl` to deploy our code. To perform the deployment, it needs `FLY_API_TOKEN` as an environment variable, which we set in the `env` section of the action definition (taking the value from the secret we have previously configured in the repository settings).

We then add this file to the Git repository, commit it, and push it. As a result, the action should run and deploy our application:

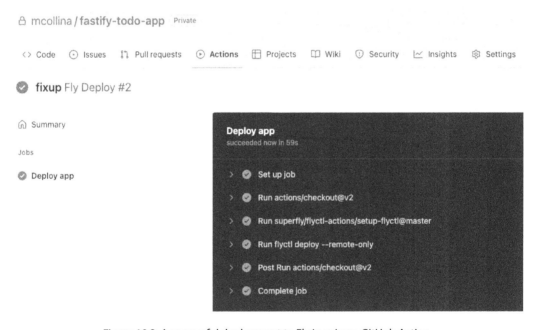

Figure 10.3: A successful deployment to Fly.io using a GitHub Action

Next, we want to configure the monitoring of our deployed application to keep it in check!

Collecting application process data

The main reason we want to collect metrics of our running Node.js process is to be able to diagnose and debug problems hours, days, or even weeks after they happened. The flow of diagnosing a production issue is described in the following figure:

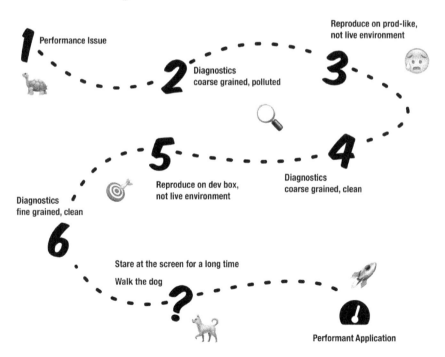

Figure 10.4: The path to solving a production issue [Source: https://mcollina.github.io/my-node-process-is-on-fire]

We need to collect data to know what happened, so we can reproduce the problem in a non-live environment – otherwise, we will not be able to understand whether we fixed the problem! After that, we use more fine-grained tools to reproduce locally and develop a fix. What tools are we going to use? We will cover them in *Chapter 13*.

Collecting the application process data with Prometheus

Prometheus is the go-to open source solution for collecting and storing time series data, and it's usually used to implement monitoring systems in applications. It's great to have statistics on how your system is performing, even under failure conditions. Prometheus also has an alert manager component that we can use to receive updates in case of outages.

Prometheus architecture is divided into different components: at its core, there is the Prometheus server, which pulls metrics from each target at precise intervals (multiple times per minute). For short-lived jobs, there is the Pushgateway, which allows Prometheus to retrieve those metrics. Visualization clients – such as Grafana – query Prometheus using PromQL.

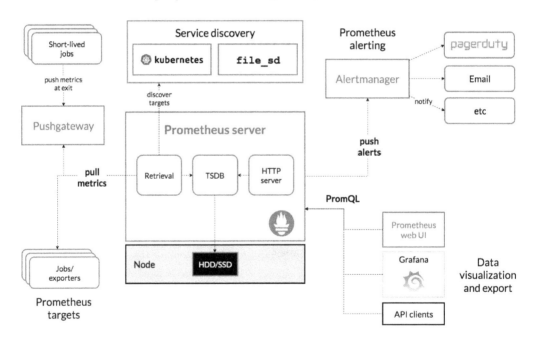

Figure 10.5: Prometheus architecture [Source: https://prometheus.io/docs/introduction/overview/]

This book is not about Prometheus and all its features; however, we will cover how to configure our Fastify application to provide data to be scraped by Prometheus. We will also cover Fly.io native support for Prometheus and native data collection capabilities.

> **Prometheus**
>
> Please refer to the Prometheus documentation for more information about Prometheus: `https://prometheus.io/docs/introduction/overview/`.

The main Node.js library that produces the metrics in the right format to be pulled by Prometheus is `prom-client` (`https://www.npmjs.com/package/prom-client`). `fastify-metrics` (`https://www.npmjs.com/package/fastify-metrics`) uses it and integrates it with Fastify.

Prometheus and Fly.io recommend that we expose the metrics on a different TCP port than our main service port, which is usually `9001`. Therefore, we want to collect the metrics for our main process and spin up a separate Fastify server to serve those metrics to the scraper. Therefore, we create a `plugins/metrics.js` file with the following content:

```
'use strict'
const fp = require('fastify-plugin')
const Fastify = require('fastify')
module.exports = fp(async function (app) {
  app.register(require('fastify-metrics'), {
    defaultMetrics: { enabled: true },
    endpoint: null,
    name: 'metrics',
    routeMetrics: { enabled: true }
  })

  const promServer = Fastify({ logger: app.log })

  promServer.route({
    url: '/metrics',
    method: 'GET',
    logLevel: 'info',
    handler: (_, reply) => {
      reply.type('text/plain')
      return app.metrics.client.register.metrics()
    }
  })
```

```
app.addHook('onClose', async (instance) => {
  await promServer.close()
})

await promServer.listen({ port: 9001, host: '0.0.0.0' })
}, { name: 'prom' })
```

Then, we need to configure Fly.io to pull our custom metrics into its monitoring solution. Thus, we add the following to `fly.toml`:

```
[metrics]
  port = 9001
  path = "/metrics"
```

Next, we run our usual trio of commands in our terminal: `git add .`, `git commit`, and `git push`. After the GitHub Action deploys our code, we can then test this by using the following:

```
$ curl https://api.fly.io/prometheus/personal/api/v1/query_
range\?step\=30 \
    --data-urlencode 'query=sum(rate(nodejs_active_handles{app="shy-
fog-346"}[5m])) by (status)' \
    -H "Authorization: Bearer YOURFLYTOKEN"
{"status":"success","isPartial":false,"data":{"resultType":
"matrix","result":[{"metric":{},"values":[[1659088433,"0"],
[1659088463,"0"],[1659088493,"0"],[1659088523,"0"],[1659088553,"0"],
[1659088583,"0"],[1659088613,"0"],[1659088643,"0"],[1659088673,"0"],
[1659088703,"0"],[1659088733,"0"]]}]}}
```

The `curl` command verifies that Fly.io is now monitoring our application – specifically, it checks the number of active handles (sockets, etc.) in Node.js. The command uses a bearer token authorization scheme (i.e., the `Authorization` header), so we will need to use the Fly.io authorization token we generated before. In case you did not copy the authorization token, you can retrieve it with `fly auth token`. When running this code, you should remember to also update the app name with yours.

Grafana will use this endpoint to fetch the data for its visualizations.

Exploring the metrics with Grafana

In this book, we will use Grafana Cloud, the hosted version of the Grafana open source project. To get started, head to `https://grafana.com/` and sign up for a new, free account. You will eventually land on the following screen:

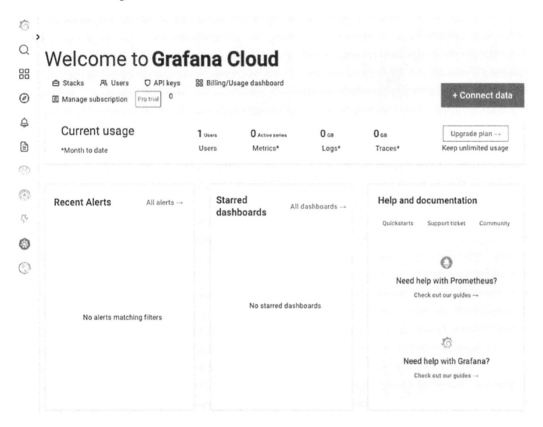

Figure 10.6: Grafana Cloud first screen

Click on **Connect data** (or in the menu, click on **Connections**), and then search for Prometheus:

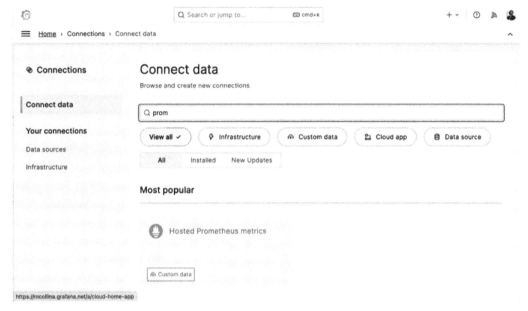

Figure 10.7: Grafana Cloud integrations and connections

Next, click on **Prometheus data source**, then **Create a Prometheus data source**:

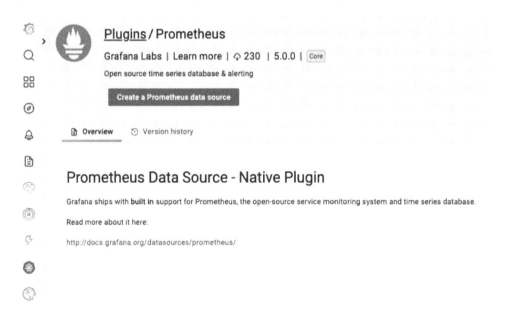

Figure 10.8: Creating a Prometheus integration in Grafana Cloud

Now, we can configure our Prometheus data source. We follow a similar approach to the `curl` command in the previous section: we must authorize Grafana to fetch data from Fly.io. To do so, we set the `Authorization` header with the `Bearer  TOKEN` value, where TOKEN is the output of `fly auth token`. The next figure shows the final configuration.

Figure 10.9: The settings of the Prometheus integration in Grafana Cloud

> **Grafana and Fly.io**
>
> You can find more information about how Fly.io can integrate with Grafana at `https://fly.io/docs/reference/metrics/#grafana`.

Now, click **Save & test**, then **Explore**. In the next screen, select `fly_instance_memory_mem_total` as a metric, apply `app = shy-fog-346` as a label (use your own app name), and click on **Run query**. The final setup will look like this:

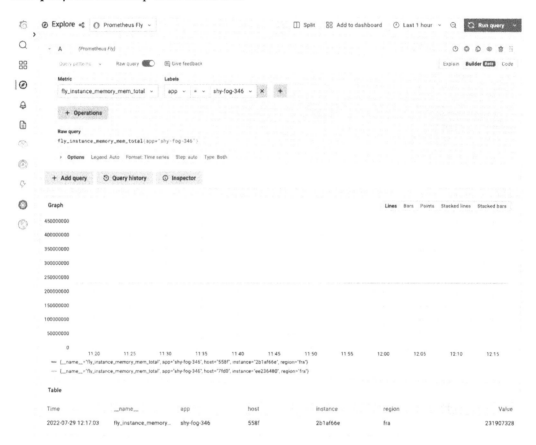

Figure 10.10: Exploring the metrics provided by Fly.io in Grafana Cloud

Here, we can see Fly.io monitoring the memory of our application.

The `fastify-metrics` module provides a few interesting metrics regarding Fastify:

- `http_request_duration_seconds_count` – the total number of HTTP requests in the interval

- `http_request_summary_seconds` – the request duration divided by percentiles

`Prom-client` provides metrics regarding the Node.js instance. You can refer to `https://github.com/siimon/prom-client/tree/721829cc593bb7da28ae009985caeeacb4b59e05/lib/metrics` for the full list of available metrics, but I recommend you to plot the following:

- `nodejs_event_loop_lag_p99_seconds` – to monitor whether we are saturating our event loop and need to further optimize our application

- `nodejs_heap_space_used_bytes` – to monitor how much memory our application uses and spot memory issues early

As a different example, we can plot the maximum latency across all routes for each instance of our application:

Figure 10.11: Plotting the latency of our server at various quantiles in Grafana Cloud

The preceding diagram shows the latency of all routes of our server at different quantiles. We are interested in the 0.95 or 0.99 quantiles because they consider most requests while excluding the outliers. This is useful to understand how our application is performing for most users.

> **Quantiles or percentiles**
>
> *"In statistics and probability, quantiles are cut points dividing the range of a probability distribution into continuous intervals with equal probabilities, or dividing the observations in a sample in the same way. There is one fewer quantile than the number of groups created. Common quantiles have special names, such as quartiles (four groups), deciles (ten groups), and percentiles (100 groups). The groups created are termed halves, thirds, quarters, etc., though sometimes the terms for the quantile are used for the groups created, rather than for the cut points."* – `https://en.wikipedia.org/wiki/Quantile`

We have learned how to integrate Fly.io with Grafana Cloud, and now we have a powerful monitoring site that we can configure to our liking. As an exercise, we recommend exploring all the metrics available in Grafana and creating a full dashboard!

> **Grafana**
>
> To learn more about Grafana, check out *Learn Grafana 7.0*, by *Eric Salituro*.

Summary

In this chapter, we have deployed our Fastify application to Fly.io with the database hosted in MongoDB Atlas. Then, we added monitoring with the Prometheus instance hosted by Fly.io. Finally, we connected Prometheus with Grafana Cloud.

In the next chapter, we will cover how to set up logging in our Fastify application to produce readable, inspectable, and useful logs.

11

Meaningful Application Logging

A piece of software does not speak to us. Sometimes we would like our application to explain what is going on and why something is not working as expected. For this reason, it is essential to teach the application how to talk to us through logs.

In this chapter, we will see how to implement meaningful application logs to help us understand what is going on in our software. It is vital to monitor that everything is working as expected, as well as to keep track of what has gone wrong.

You will learn how to set up the perfect logging configuration without losing essential information. Moreover, you will discover how to avoid logging sensible data and print out only what you need.

The learning path we will cover in this chapter is as follows:

- How to use Fastify's logger
- Enhancing the default logger
- Collecting the logs
- Managing distributed logs

Technical requirements

As in the previous chapters, you will need the following:

- A working Node.js 18 installation
- The VS Code IDE from `https://code.visualstudio.com/`
- A public GitHub repository
- A working command shell

All the snippets in this chapter are on GitHub at `https://github.com/PacktPublishing/Accelerating-Server-Side-Development-with-Fastify/tree/main/Chapter%2011`.

How to use Fastify's logger

Application logging is one of the pillars of application observability. It provides useful information to understand what the system is doing and narrow down where the bug arises. A good logging setup has the following qualities:

- **Consolidate**: This sends the logs to a designated output.
- **Structure**: This provides a format to query and analyze over text messages.
- **Context**: This adds metadata based on the execution environment.
- **Security**: This applies filter and data redaction to hide sensitive information.
- **Verbosity**: This sets the log severity based on the environment. Unfortunately, logging is not free, as it costs the system's resources, so we can't just log everything without impacting the application's performance.

Generally, you can easily understand whether application logs are implemented correctly or not. But you should never underestimate the importance of logging.

In fact, when you are having trouble in production caused by an unknown and mysterious error, and the error log prints out a vague message such as `undefined is not a function`, at that moment, it is already too late to act, and you can't reproduce the error event because it's gone.

The logger is not just a life-saver when things are burning down. The log is an enabler in accomplishing the following:

- **Auditing**: This tracks events to recreate the user's navigation and API usage
- **Metrics**: This traces the time spent per request or operation to identify possible bottlenecks
- **Analysis**: This counts the endpoint metrics to plan code base refactoring and optimizations, such as choosing routes that slow down the system and need to be isolated
- **Alerting**: This triggers a warning when specific events are recognized by monitoring the metrics

Fastify helps you manage all kinds of situations because it has all the tools you need to build a production-ready application. As introduced in *Chapter 1*, the Fastify framework includes the `pino` logger module (`https://getpino.io`), which provides all you need to let your application speak to you. We need to understand some basic concepts before digging into the code, or we could get lost in the configuration maze.

How Pino logs the application's messages

Before starting, we need a brief introduction to Pino, the fastest logger in the Node.js ecosystem. By default, it prints out to the **stdout** and **stderr** streams. These streams print out the string message to our shell's console to let us read the output on the screen. You can go into further detail by reading this article: `https://en.wikipedia.org/wiki/Standard_streams`.

To log something, we need to call one of the methods corresponding to a **log level**. Here is the list in ascending severity order:

- `log.trace()`: This is used to print very low-severity messages
- `log.debug()`: This is the level for low severity diagnostic messages
- `log.info()`: This prints logs for informational text
- `log.warn()`: This indicates problems or cases that should not have occurred
- `log.error()`: This is used to trace failures
- `log.fatal()`: This indicates a major error that could cause the server to crash

You can configure the log to turn the log's output on or off based on the log level. For example, by setting `warn` as the threshold, all the lower severity logs will not be shown.

All the log functions share the same usage interface. To clarify, we can recap the interface into these forms:

```
const pino = require('pino')
const log = pino({ level: 'debug' })
log.debug('Hello world') // [1]
// {"level":30,"time":1644055693539,"msg":"Hello world"}
log.info('Hello world from %s', 'Fastify') // [2]
// {"level":30,"time":1644055693539,"msg":"Hello world from
    Fastify"}
log.info({ hello: 'world' }, 'Cheers') // [3]
// {"level":30,"time":1644055693539,"hello":"world","msg":
    "Cheers"}
log.info({ hello: 'world' }, 'Cheers %s', 'me') // [4]
// {"level":30,"time":1644055693539,"hello":"world","msg":
    "Cheers me"}
```

The first syntax, [1], logs a simple string message. It helps debug and trace.

The second interface, [2], shows us how Pino interpolates strings: when the first string argument is found, all the subsequent parameters will be used to fill its placeholder. In the [2] code example, we are using `%s` for the string placeholder: you can use `%d` for numbers and `%o` for JSON objects.

> **String interpolation optimization**
>
> When you plan to hide the log level, you should prefer the `pino` string interpolation over template literals or string concatenation. To give you an example, writing `log.debug(`Hello ${name}`)` means that every time the statement is executed, the template's literal is processed, even if the DEBUG log level is hidden. This is a waste of resources. Instead, the `log.debug('Hello $s', name)` statement does not execute the string interpolation if the corresponding log level does not need to be printed.

The third [3] log syntax shows that we can include a JSON object in the log output by setting a string message. This API can be used to log `Error` objects too. Finally, [4], which we saw in the previous code snippet, sums up the JSON object log and the string interpolation.

As you have seen, the output is a JSON logline string (`https://jsonlines.org/`) that adds additional information by default, such as the log time in milliseconds since the Unix timestamp, and the log level as an integer, and the system's coordinates (which have been removed from the snippet output for compactness). You can customize every JSON field printed out by managing the options or implementing a **serializer**. The log serializer is a function that must return a JSONifiable object, such as a string or a JSON object. This feature lets us add or remove information from the logline, so let's see an example:

```
const pino = require('pino')
const log = pino({
  serializers: {
    user: function userSerializer (value) {
      return { id: value.userId }
    }
  }
})
const userObj = { userId: 42, imageBase64: 'FOOOO...' }
log.info({ user: userObj, action: 'login' })
// {"level":30,"time":1644136926862,"user":{"id":42},
    "action":"login"}
```

In the previous code snippet, `userObj` contains the `imageBase64` property that must not be logged because it is not useful, and it is a huge `base64` string that can slow down the logging process. So, when we need to log an application's account, we can define the convention so that the corresponding JSON must be assigned to a `user` property as shown. Since we have defined a `user` serializer when we declared the log instance, the `userSerializer` function will be executed, and its output will be used as output. We have centralized the logic to log a `user` object into a single operation by doing so. Now, we can log the user in the same way from the whole application.

The JSON format output can't be changed out of the box by design. This is a requirement of being the fastest logger in the Node.js panorama. To be flexible for each developer, the ecosystem of `pino` is supplied with a lot of **transporters**. These are plugins that let you do the following:

- **Transform**: This transforms the log output into another string format different from the JSON one, such as `syslog` (`https://datatracker.ietf.org/doc/html/rfc5424`)

- **Transmitting**: This transmits the logs to an external system, such as **log management software**

The JSON format is broadly supported and can be easily stored and processed by external analysis tools called log management software. This software allows you to collect the data, search and report, and store it for retention and analysis.

The transport's configuration depends on what you are going to do within the logs. For example, if you already have the log management software in place, you may need to adapt to its logs string format.

> **Using pino as a logger**
>
> Note that there are multiple ways to transmit the records to a third-party system that do not involve `pino`, such as a platform command, but this goes beyond the main topic of this book.

Now that you have an overview of `pino`, we need to go into the logging configuration details exposed by Fastify. So, let's jump in to appreciate all the aspects of customizing logging configuration.

Customizing logging configuration

In *Chapter 3*, we saw how to set the log level and the pretty print configuration but there is much more than that to customizing the logger.

The main log's configuration must be provided to the Fastify instance during its initialization. There are three ways to do that, as depicted by the following code snippet:

```
const app = fastify({ logger: true }) // [1]
const app = fastify({
  logger: pinoConfigObject // [2]
})
const app = fastify({
  logger: new MyLogger() // [3]
})
```

The first option, **[1]**, is good when we need to try some code snippets and need a quick and dirty way to see the default Fastify's log configuration.

The second interface, [2], is the one we have used so far in the book, last mentioned in *Chapter 6*. The `logger` server's property accepts a JSON object corresponding to the `pino` options. We set those options by configuring the `fastify-cli -l info --pretty-logs` arguments. You can get the full options list by reading the Pino documentation at `https://getpino.io/#/docs/api?id=options-object`.

The last input, [3], allows you to provide a new logger module and not use `pino` by default instead. It enables you to provide a custom `pino` instance too. This is useful if you want to use a major new release without waiting for a new Fastify release.

> **Don't change the logger library**
>
> In this chapter, you will not learn how to change Fastify's logger module. We hope you understand why you should not do it, otherwise, you will lose all the fine configurations the framework gives you. If you need to do it, you can read our reasoning on Stack Overflow: `https://stackoverflow.com/questions/55264854/how-can-i-use-custom-logger-in-fastify/55266062`.

These options let us customize the log's baseline settings. The possibilities are not over yet. You can do a lot more using Fastify! You can customize the log level and the serializers at the following levels:

- Server instance level
- Plugin level
- Route level

Let's see an example. First, we need a simple handler that prints a `hello` array:

```
async function helloHandler (request, reply) {
  const hello = ['hello', 'world']
  request.log.debug({ hello })
  return 'done'
}
```

Now we can reuse this handler to see the different behaviors in the three cases listed previously:

```
const app = fastify({
  logger: {
    level: 'error',
    serializers: {
      hello: function serializeHello (data) {
        return data.join(',')
      }
    }
  }
})
```

```
})
app.get('/root', helloHandler)
app.inject('/root')
```

The expected output is none because the `hello` log is at debug level, and we set an `error` threshold. The `serializeHello` function is not executed at all. This setting will be the default for every child context and route since it is assigned to Fastify's server instance root.

As mentioned earlier in this section, we can overload the server's default log configuration by using two additional plugin options managed by Fastify itself:

```
app.register(async function plugin (instance, opts) {
  instance.get('/plugin', helloHandler)
}, {
  logLevel: 'trace',
  logSerializers: {
    hello: function serializeHello (data) {
      return data.join(':')
    }
  }
})
app.inject('/plugin')
```

The `logLevel` and `logSerializers` properties are handled by Fastify, and they will overwrite the default log's setting for the `plugin` instance. This means that even the plugin's child contexts and routes will inherit the new log configuration. In this case, the expected output is a string concatenated by a double colon:

```
{"level":20,"time":1644139826692,"pid":80527,"hostname":"MyPC","reqId"
:"req-1","hello":"hello:world"}
```

The route option object supports the same special `logLevel` and `logSerializers` properties:

```
app.get('/route', {
  handler: helloHandler,
  logLevel: 'debug',
  logSerializers: {
    hello: function toString (data) {
      return data.join('+')
    }
  }
})
app.inject('/route')
```

In the route code example, the output expected is now a string concatenated by the plus symbol:

```
{"level":20,"time":1644140376244,"pid":82198,"hostname":"MyPC
","reqId":"req-1","hello":"hello+world"}
```

This fine granular log-level setup is a great feature. For example, you can set an `error` level to stable routes and `info` thresholds for brand new routes that require beta usage before considering them durable. This optimization reduces the beta endpoints' impact on the system, which may suffer a low log level for every application's route, causing a considerable increment on the logline counter.

> **Logs are not free**
>
> As we said previously, logs are not free: they cost the system's resource utilization, but there is also an economic cost involved. Some log management systems apply pricing per logline or logline size. Fastify helps you control this project management aspect.

Well done, you have now seen how to tweak your application's logger to your needs. So far, we have built the basis to start from in order to customize Pino. In the next section, we will use all the options we have learned about in this chapter so far.

Enhancing the default logger configuration

You know the options available in the logger, but what is a valuable configuration, and how can we integrate it into our application? First of all, we need to define which logs we expect for every request:

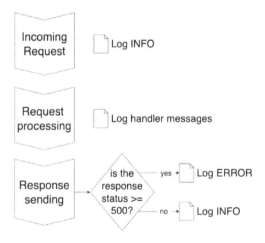

Figure 11.1 – Request logs logic

As shown in *Figure 11.1*, we expect these log lines for each request: one for the incoming request, optionally, how many handler's log messages you implement, and finally, one for the response output. All these log lines will be connected to each other by the **reqId** (request-id) field. It is a unique identifier generated for each request whenever the server receives it. We already described the reqId property in *Chapter 1* in the *The Request component* section.

We can start implementing this by doing the following:

- Exploiting the default request and response log implemented by Fastify

- Executing a custom log statement by using the onRequest and onSend hooks

The first option is more straightforward to implement, but it has less configurability since you can't customize the message. Moreover, Fastify's default request log runs before any other hooks, and you will not be able to read any additional information. The latter option is easy to set up and is open to advanced configurations.

We will follow the custom implementation for our *Fastify to-do list* application implemented till the previous *Chapter 10* and follow the logic in *Figure 11.1* by doing the following:

1. Turning off the Fastify default request log. To do that, edit the configs/server-options. js file, by adding the disableRequestLogging: true property we saw in *Chapter 1*.

2. Adding an onRequest and onResponse hook in the plugins/error-handler.js file.

The hook will look like the following:

```
fastify.addHook('onRequest', async (req) => {
  req.log.info({ req }, 'incoming request')
})
fastify.addHook('onResponse', async (req, res) => {
  req.log.info({ req, res }, 'request completed')
})
```

Note that we have added the req object to the response log, as it is useful to add the request's URL information. The rationale for adding these logging hooks in the error-handler file is the centralization of the HTTP request tracing. This file now contains all the minimal request-logging processes.

If we compare Fastify's default output within our setup, we lose the responseTime property from the response's log line. Now, we can fix this by adding the first custom log serializer. Let's create a new configs/logger-options.js file. This document will contain Pino's options as follows:

```
module.exports = {
  level: process.env.LOG_LEVEL,
  serializers: {
    res: function (reply) {
      return {
```

```
            statusCode: reply.statusCode,
            responseTime: reply.getResponseTime()
          }
        }
      }
    }
```

It is possible to add the HTTP request duration in the log output in order to define the `res` serializer. Note that it is necessary to set the `level` property first. In fact, setting the logger configuration has priority over the `fastify-cli` arguments, so adding a new LOG_LEVEL environment variable will be mandatory. We must also remember to register this new variable into the `schemas/dotenv.json` file.

To load the new `logger-options.js` file into the `configs/server-options.js` file correctly, do the following:

```
const loggerOptions = require('./logger-options')
module.exports = {
  disableRequestLogging: true,
  logger: loggerOptions,
  ...
```

Now, calling an endpoint will produce the two log lines that we can format to improve the readability. The `onRequest` hook will print the following:

```
{
  "level": 30,
  "time": 1644273077696,
  "pid": 20340,
  "hostname": "MyPC",
  "reqId": "req-1",
  "req": {
    "method": "GET",
    "url": "/",
    "hostname": "localhost:3001",
    "remoteAddress": "127.0.0.1",
    "remotePort": 51935
  },
  "msg": "incoming request"
}
```

The `onResponse` hook prints the response information:

```
{
  "level": 30,
  "time": 1644273077703,
  "pid": 20340,
  "hostname": "MyPC",
  "reqId": "req-1",
  "req": {
    "method": "GET",
    "url": "/",
    "hostname": "localhost:3001",
    "remoteAddress": "127.0.0.1",
    "remotePort": 51935
  },
  "res": {
    "statusCode": 200,
    "responseTime": 8.60491693019867
  },
  "msg": "request completed"
}
```

You can customize the tracing message's properties by adding the data you need. As mentioned in the *How to use Fastify's logger* section, a good logline must have a Context.

A request's Context helps us answer the following common questions:

- Which endpoint has produced the log?

- Was the client logged in and who was the user?

- What was the request's data?

- Where are all the log lines related to a specific request?

Now, we can implement the answers to all these questions by enhancing the log's serializers. So let's go back to the `configs/logger-options.js` file and create a new `req` serializer for the HTTP request object:

```
serializers: {
  req: function (request) {
    const shouldLogBody = request.context.config.logBody
    === true
    return {
      method: request.method,
      url: request.url,
      routeUrl: request.routerPath,
      version: request.headers?.['accept-version'],
```

```
      user: request.user?.id,
      headers: request.headers,
      body: shouldLogBody ? request.body : undefined,
      hostname: request.hostname,
      remoteAddress: request.ip,
      remotePort: request.socket?.remotePort
    }
  },
  res: function (reply) { ...
```

The new `req` configuration emulates the default one and adds more helpful information:

- The `routeUrl` property seems redundant, but it prints the route's URL. This is useful for those URLs that contain path parameters: in this case, the `url` log entry includes the input values set by the client. So, for example, you will get `url: '/foo/42'`, `routeUrl: '/foo/:id'`.

- The `user` field shows who made the HTTP request and whether the client had a valid login. Since the login process happens at different moments, this property is expected only when the `reply` object is logged.

- The `headers` occurrence outputs the request's headers. Note that this is an example. It would be best to define what headers you need to log first and then print out only a limited set.

- The `body` property logs the request's payload. The first thing to note is that logging all the request's body could be harmful due to the size of the body itself or due to sensible data content. It is possible to configure the routes that must log the payload by setting a custom field into the route's options object:

```
fastify.post('/', {
  config: { logBody: true },
  handler: async function testLog (request, reply) {
    return { root: true }
  }
})
```

Another detail you need to pay attention to is the `||` `undefined` condition to avoid falsy values in the logline. `pino` evicts all the `undefined` properties from the log record, but it prints the `null` ones.

We have added more and more context to the request's logline, which helps us to collect all the information we need to debug an error response or simply track the application's routes usage.

Printing the context into the logs may expose sensitive information, such as passwords, access tokens, or personal data (email, mobile numbers, and so on). Log-sensitive data should not happen, but let's see how you can hide this data in the next section.

How to hide sensitive data

In the previous section, we printed out a lot of information, but how can we protect this data? We mentioned that a good logline must provide security. Therefore, we must execute **data redaction**, which masks the strings before they are logged. `pino` supports this feature out of the box, so it is necessary to tweak the `configs/logger-options.js` file once more:

```
module.exports = {
  level: process.env.LOG_LEVEL || 'warn',
  redact: {
    censor: '***',
    paths: [
      'req.headers.authorization',
      'req.body.password',
      'req.body.email'
    ]
  },
  serializers: { …
```

The `redact` option lets us customize which lookup paths should be masked within the `censor` string to use in place of the original value. We set three direct lookups in the code example. Let's examine `req.body.password`: if the logger finds a `req` object within a `body` entity and a `password` property, it will apply the redaction.

The `redact` configuration option in action will hide the property's value from the log as follows:

```
{
  ...
  "req": {
    "method": "POST",
    "url": "/login",
    "routeUrl": "/login",
    "headers": {
      "authorization": "***",
  "   "content-length": "60"
    },
    "body": {
      "email": "***",
      "password": "***"
    },
    ...
  },
  "res": {
    "statusCode": 200,
    "responseTime": 11.697166919708252
```

```
    },
    "msg": "request completed"
}
```

The log redaction is a mandatory step to provide a secure system and avoid potential data leaks. You need to have a clear idea about which routes will log the user's input. The logBody route option we have seen in the previous section helps you control it and act accordingly.

Note that the redaction will not be executed if the password field is moved into a wrapper object. It won't work if the property changes its case to Password (capital letter). The redact supports wildcards such as *.password, but it also heavily impacts the logger's performance and, consequently, the application itself. You can gather further information on this aspect by reading the official documentation at https://github.com/pinojs/pino/blob/master/docs/redaction.md#path-syntax.

To be sure that a detailed configuration, like the one we're discussing, is running properly, we can write a test that checks the correctness censorship configuration. We can create a test/logger.test.js file to cover this case. We are going to redirect the application's log messages to a custom stream, where we will be able to process each log line. To do so, we must run npm install split2 to ease the stream handling. Then, we can implement the following test:

```
const split = require('split2')
t.test('logger must redact sensible data', async (t) => {
  t.plan(2)
  const stream = split(JSON.parse)
  const app = await buildApp(t,
    { LOG_LEVEL: 'info' },
    { logger: { stream } }
  )
  await app.inject({
    method: 'POST',
    url: '/login',
    payload: { username: 'test', password: 'icanpass' }
  })
    for await (const line of stream) {
    // [1]
  }
})
```

The code snippet is a scaffolding test that reads all the log lines that our application emits. We provide the `logger.stream` option to redirect the output of `pino` without modifying the `configs/logger-options.js` options, such as the `reduct` option. After the HTTP request injection, we can check the emitted logs. The assertions in [1] will look as follows:

```
if (line.msg === 'request completed') {
  t.ok(line.req.body, 'the request does log the body')
  t.equal(line.req.body.password, '***', 'field
  redacted')
  break
}
```

Whether the test we just created passes or not depends on your `.env` file settings. This is caused by an error during the loading phase. If you try to think about the application's files load sequence when you run the tests, you will get the following steps:

1. The `.env` file is loaded by the `fastify-cli` plugin.

2. The `fastify-cli` plugin loads the `configs/server-options.js` file.

3. The `configs/logger-options.js` file is loaded to build the Fastify instance. It will use the current `process.env.LOG_LEVEL` environment variable.

4. Still, the `fastify-cli` plugin merges the loaded configuration with the third `buildApp` parameter, where we set the `stream` option.

5. The `@fastify/autoload` plugin will load all the applications, including the `plugins/config.js` file, which will read the test's `configData` property that we saw in *Chapter 9*.

How can we fix this situation? Luckily, `pino` supports the log level set at runtime. So we need to update the `plugins/config.js` file, adding this statement:

```
module.exports = fp(async function configLoader (fastify, opts) {
  await fastify.register(fastifyEnv, { ... })
  fastify.log.level = fastify.secrets.LOG_LEVEL
  // ...
```

It is possible to update the log level at runtime, and it gives us the control to customize the log level from the tests without impacting the production scenario. Moreover, we have learned that we can change the log severity at runtime by changing the verbosity in the *How to use Fastify's logger* section! This enables you to adapt the log to situations, such as logging for a fixed amount of time in the `debug` mode and then back to `warn` automatically without restarting the server. The possibilities are truly endless.

Now you know the best setup to print out a useful context and provide security to every single logline. Now, let's find out how to store them.

Collecting the logs

Reading the logs can be done on your PC during development, but it is unrealistic to carry it out during production or even in a shared test environment. Reading the records is not scalable, but if you try to estimate the number of logs your application will write at the info level, you will get an idea.

Having 10 clients send 5 requests per second is equal to 100 lines per second per day. Therefore, the log file would be more than 8 million rows, just for a single application installation. If we scale the application to two nodes, we need to search in two different files, but if we followed this path, the log files would be useless because they would be inaccessible.

As mentioned at the beginning of this chapter, a good log setup allows us to consolidate to a log management software destination, so let's see how to design it.

How to consolidate the logs

Before considering where to consolidate the logs, we must focus on the actor who submits the logs to the destination. All the most famous log management software systems expose an HTTP endpoint to submit the records to, so the question is, should our application submit the log to an external system?

The answer is *no*, and we will learn why starting with the following diagram:

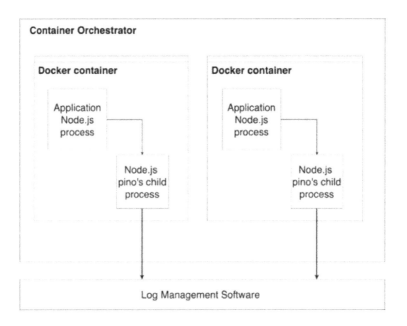

Figure 11.2 – Node.js process submits the logs

In the scenario in *Figure 11.2*, we can see a hypothetical production configuration where our application, packed into a Docker container, sends the log output directly to the log management software. pino transporters are capable of spinning up a new Node.js child process to reduce the application's cost of logging to a minimum. Now that we know more about this excellent optimization, we would still like to pose the following questions:

- What if the log management software is offline due to maintenance or a server issue?
- Should the application start if it can't connect to the external log destination?
- Should logging impact the application performance?
- Should the application be aware of the logs' destination?

We think you will agree that the only possible answer to all these questions is *no*. Note that this scenario applies to all the system architectures because *Figure 11.2* shows a Docker container that runs our application. Still, it could be a server that runs our Node.js application manually.

For this reason, the right action to implement the log consolidation is a two-step process, as shown in the following corrected schema:

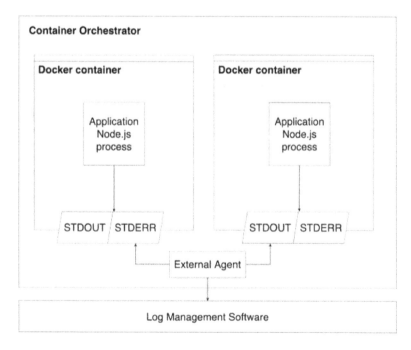

Figure 11.3 – External agent submits the logs

In the architecture shown in *Figure 11.3*, you will notice that the application should log to the `stdout` and `stderr` output streams. By doing this, you don't need to configure a `pino` transport, and the container's host will use fewer resources. The container orchestrator will be responsible for reading and processing the messages. This task is carried out by a software **agent**, which is usually provided by the log management software. Within this schema, we decoupled the log application logic from the log's destination, isolating the former by whatever issue the external store may face.

Decoupling the log message producer from the message consumer is not 100% accurate: your application should generate a supported structure through the log management software, so you need to verify this with the carrier you choose.

For example, some external log stores require a `@timestamp` field instead of Fastify's default `time` property, but it can be easily configured by tweaking the `timestamp` option of `pino`. For completeness, here is an example of editing the `configs/logger-options.js` file:

```
module.exports = {
  level: process.env.LOG_LEVEL,
  timestamp: () => {
    const dateString = new Date(Date.now()).toISOString()
    return `,"@timestamp":"${dateString}"`
  },
  redact: { ... }
```

The previous code example shows that it is possible to customize the output date format too, replacing the default epoch time with the number of milliseconds since January 1st, 1970 at 00:00 GMT.

Nevertheless, you will need to configure a log retention policy. The logs may not be helpful after a reasonable amount of time, or you may need to archive the messages for a longer amount of time to accomplish some legal agreement. Good log management software helps you solve all these trivial tasks.

On the other hand, you may face the need to submit your application's log directly to an external service. You might not have a container orchestrator at your disposal or want to log to the file. So, let's see how to deal with this use case.

Consolidating logs by using Pino transports

Pino transports are the components designed to submit the logs to a destination, such as a filesystem or log management software. Every transport is a Node.js **worker thread**. A worker thread enables you to execute JavaScript in parallel and is more valuable when it performs CPU-intensive operations. By doing this, `pino` saves the main application from running trivial code to manage the logs.

As usual, you can configure it by setting the `configs/logger-options.js` file. The first experiment is to log in to the filesystem, so let's have a look at the following code:

```
module.exports = {
  level: process.env.LOG_LEVEL,
```

```
transport: {
  target: 'pino/file',
  options: {
    destination: require('path').join(__dirname,
    '../logs/errors.log')
  },
  level: 'error'
},
timestamp: () => { ...
```

While adding the `transport` configuration, we must set some parameters:

- `target` is the transport to execute. You can define a path to run a custom transport implementation or a plugin named `pino-redis` or `pino-mongodb` to forward the logs to the corresponding databases.

- The `options` object is a parameter provided to the transport implementation.

- Optionally, a `level` property limits the logs that the transporter will receive and process.

This simple setup will store all the `error` logs in a dedicated file.

> **Log rotation**
>
> It is important to mention that `pino/file` does not take care of log rotation or filesystem exhaustion. For these kinds of tasks, you can rely on tools such as `logrotate` at `https://github.com/logrotate/logrotate`.

As you will see by running the Fastify server, this setup will not print any logs to `stdout`, making the development process very hard. So we need to configure two means of transport for our use case. To do this, we need to edit the previous configuration a bit:

```
module.exports = {
  level: process.env.LOG_LEVEL,
  transport: {
    targets: [
      {
        target: 'pino/file',
        options: {
          destination: require('path').join(__dirname,
          '../logs/error.log')
        },
        level: 'trace'
      },
      {
        target: 'pino/file',
```

```
        options: { destination: 1 }
      }
    ]
  },
  timestamp: () => { ...
```

We can list all the transports we need by adding a `targets` array property. In this case, first `target` is like the one configured earlier at the beginning of this section to store the messages into a file. The second one is still a `pino/file` target. The `pino/file` target is a built-in transport that writes the logs to a file descriptor. If the `destination` argument is a string, a file descriptor will be opened, as shown in the first target's item. Instead, if the parameter is an integer, the transport will write to it. By default, every Unix process has three standard file descriptors:

- `0` `stdin`: This is for the standard input
- `1` `stdout`: This is for the standard output
- `2` `stderr`: This is for the standard error

If you would like to read more about this, you can start with this article: `https://en.wikipedia.org/wiki/File_descriptor`.

When a transport configuration does not contain the `level` parameter, it will receive all the messages based on the primary logger instance's severity setup.

For a complete transport list, you can rely on the official documentation: `https://github.com/pinojs/pino/blob/master/docs/ecosystem.md`. You can find inspiration or integrations already developed by the `pino` maintainers to help you find a good log management software for use in your daily work.

Log conflicts

By using the transports in your application, you may have issues running the `test/logger.test.js` file. Transports have priority over the `stream` parameter we set in the *How to hide sensitive data* section, so the test stream will not receive the log lines. To fix this, you need to omit the `transports` options when you run the tests. You can do that by checking the `process.env` object.

This section has discussed log collecting challenges and how to deal with them by creating a solid system architecture or a single configuration of `pino`. You are now aware of these topics, which are often overlooked. Now that we have a single point of access to read the application's logs, we need to know how to make them searchable.

Managing distributed logs

Fastify has simplified our job many times, and it always has an excellent way to complete complex tasks. This is the case for distributed logs. By distributed records, we mean the situation of a single client's HTTP request that leads to multiple requests across our application or the whole system. In the previous *Enhancing the default logger configuration* section, we learned how to add context to the Fastify logs, but how can we connect all those messages to a single HTTP request? And what about an external API call to another Fastify server? To do this, we must configure the `reqId` request-id properly.

The request-id is the identifier across your entire system to be able to trace all the logs generated by a single HTTP request. In Fastify, you can't easily remove the request-id field from the output message, so whenever you use `request.log` or `reply.log`, you will get the `reqId` property.

We can customize the request-id field name and value generation. As you have seen, the default format is `reqId: "req-${counter}"`, where the counter is an in-memory numeric sequence starting from 1 to ~2 billion.

To modify those two values, we need to update the `configs/server-options.js` file:

```
module.exports = {
  disableRequestLogging: true,
  logger: loggerOptions,
  requestIdLogLabel: 'requestId',
  requestIdHeader: 'x-request-id',
  genReqId (req) {
    return req.headers['x-amz-request-id'] ||
    crypto.randomUUID()
  },
  ajv: { ...
```

These new log options are not included in the `logger` property because they are under Fastify's domain. The `requestIdLogLabel` parameter changes the output field name in the logline.

The `requestIdHeader` field accepts a string that represents an HTTP request's header name. This parameter will always be evaluated to set the incoming HTTP ID. You can disable it by setting the value to `false`. The `genReqId` function is the final fallback when a valid header is not found, or the header's check has been disabled. The function's implementation is up to you: we access another request's header in the previous code example. We will generate a random **universally unique identifier (UUID)** string to assign to the incoming call if it is missing.

Finally, we must remember to prefer `request.log` or `reply.log` over the `fastify.log` object to get the request-id in the log's output. Moreover, we need to propagate the request-id to any internal calls. For example if the service `POST/register`, created in *Chapter 8*, notifies the billing microservice, we need to include the request id value in the correct header:

```
async function fastifyHandler (request, reply) {
  const userId = await saveUser(request.body)
  await billingService.createWallet(userId, {
    headers: {
      'x-request-id': request.id
    }
  })
  return { done: true }
}
```

The code snippet shows you a possible example of how to propagate the request-id to keep all the client's calls related to their originator request.

By setting up Fastify's request-id, you will be able to filter your logs by using the log management software that collects the application's logs. An error log will easily show its entire flow into your system, allowing you to debug and dig deeper to solve the issues effectively.

The request-id log options were the last configuration we needed to discuss to complete our journey into the world of logs. You have acquired all the knowledge needed to set and customize all these properties to better fit your needs, and this brings us to the end of this chapter.

Summary

In this chapter, we have discussed the importance of logging and how to focus on improving this feature. We have explored many new Fastify options to enhance the readability of the application's logs. You now have a complete overview of the `pino` module and know how to customize it based on your system's architecture.

We have successfully configured the log's redaction to secure our log files and hide sensitive data, which is one of the hardest things to do, but we still made it easy, thanks to Fastify's ready-to-use toolchain in production.

You have read about some architectural patterns to collect the logs in a complex scenario and the aspects you need to consider when dealing with logs.

You are ready to proceed to *Chapter 12*, where you will learn how to split our application into smaller and independent pieces.

Part 3: Advanced Topics

After covering the basics of the framework and developing a full-fledged, real-world, production-grade application, it's time to move on and see some advanced scenarios. Fastify is flexible enough to run on both your development machine and internet-scale infrastructure. The chapters in this part will touch on distinct auto-conclusive topics that might be crucial to your application's success!

In this part, we cover the following chapters:

- *Chapter 12, From a Monolith to Microservices*
- *Chapter 13, Performance Assessment and Improvement*
- *Chapter 14, Developing a GraphQL API*
- *Chapter 15, Type-Safe Fastify*

12
From a Monolith to Microservices

Our little application has started gaining traction, and the business has asked us to redesign our API while keeping the previous version running to ease the transition. So far, we have implemented a "monolith" – all of our application is deployed as a single item. Our team is very busy with evolutive maintenance, which we cannot defer. Then, our management has a "Eureka!" moment: let's add more staff.

Most engineering management books recommend that the team size should never increase beyond eight people – or if you're Amazon, no larger than the amount that can share two large pizzas (that's Jeff Bezos's two-pizza rule). The reason is that with over eight people, the number of interconnections between team members grows exponentially, making collaboration impossible. An often overlooked solution is to not grow the team, but rather slow down delivery.

A solution to our growing pains could be to split our team in two. This is a bad idea because having two teams working at the same time on the same code base would only cause them to step on each other toes. Unfortunately, this is unlikely to happen, as the demand for digital solutions grows yearly. What should we do?

The first step is to structure our monolith into multiple modules to minimize the chance of conflict between different teams. Then, we can split it into microservices so that teams could be in charge of their deployments. Microservices are powerful only if we can arrange the software architecture with the team boundaries in mind.

In this chapter, we will start by making our monolith more modular, and then we will investigate how to add new routes without increasing our project's complexity. After that, we will split the monolith and use an API gateway to route the relevant calls.

Ultimately, we will cover all the operator questions: logging, monitoring, and error handling.

So, in this chapter, we will cover the following topics:

- Implementing API versioning
- Splitting the monolith
- Exposing our microservice via an API gateway
- Implementing distributed logging

Technical requirements

As mentioned in the previous chapters, you will need the following:

- A working Node.js 18 installation
- A text editor to try the example code
- Docker
- An HTTP client to test out code, such as CURL or Postman
- A GitHub account

All the snippets in this chapter are on GitHub at `https://github.com/Packt Publishing/Accelerating-Server-Side-Development-with-Fastify/tree/ main/Chapter%2012`.

Implementing API versioning

Fastify provides two mechanisms for supporting multiple versions of the same APIs:

- Version constraints are based on the `Accept-Version` HTTP header (`https://www. fastify.io/docs/latest/Reference/Routes/#constraints`)
- URL prefixes are simpler and my go-to choice (`https://www.fastify.io/docs/ latest/Reference/Routes/#route-prefixing`)

In this section, we will cover how to apply both techniques, starting with the following base server:

```
// server.js
const fastify = require('fastify')()

fastify.get('/posts', async (request, reply) => {
  return [{ id: 1, title: 'Hello World' }]
})
```

```
fastify.listen({ port: 3000 })
```

Version constraints

The Fastify constraint system allows us to expose multiple routes on the same URL by discriminating by HTTP header. It's an advanced methodology that changes how the user must call our API: we must specify an `Accept-Version` header containing a semantic versioning pattern.

For our routes to be version-aware, we must add a `constraints: { version: '2.0.0' }` option to our route definitions, like so:

```
const fastify = require('fastify')()
async function getAllPosts () {
  // Call a database or something
  return [{ id: 1, title: 'Hello World' }]
}

fastify.get('/posts', {
  constraints: { version: '1.0.0' }
}, getAllPosts)

fastify.get('/posts', {
  constraints: { version: '2.0.0' }
}, async () => { posts: await getAllPosts() })

app.listen({ port: 3000 })
```

We can invoke our v1.0.0 API with the following:

```
$ curl -H 'Accept-Version: 1.x' http://127.0.0.1:3000/posts
[{"id":1,"title":"Hello World"}]
```

We can invoke our v2.0.0 API with the following:

```
$ curl -H 'Accept-Version: 2.x' http://127.0.0.1:3000/posts
{"posts":[{"id":1,"title":"Hello World"}]}
```

Invoking the API without the `Accept-Version` header will result in a 404 error, which you can verify like so:

```
$ curl http://127.0.0.1:3000/posts
{"message":"Route GET:/posts not found","error":"Not
Found","statusCode":404}
```

As you can see, if the request does not have the `Accept-Version` header, a 404 error will be returned. Given that most users are not familiar with `Accept-Version`, we recommend using prefixes instead.

URL prefixes

URL prefixes are simple to implement via encapsulation (see *Chapter 2*). As you might recall, we can add a `prefix` option when registering a plugin, and Fastify encapsulation logic will guarantee that all routes defined within the plugins will have the given prefix. We can leverage prefixes to structure our code logically so that different parts of our applications are encapsulated.

Let's consider the following example in which each file is separated by a code comment:

```
// server.js
const fastify = require('fastify')()
fastify.register(require('./services/posts'))
fastify.register(require('./routes/v1/posts'), { prefix: '/v1'
})
fastify.register(require('./routes/v2/posts'), { prefix: '/v2'
})
fastify.listen({ port: 3000 })

// services/posts.js
const fp = require('fastify-plugin')
module.exports = fp(async function (app) {
  app.decorate('posts', {
    async getAll () { // Call a database or something
      return [{ id: 1, title: 'Hello World' }]
    }
  })
})

// routes/v1/posts.js
```

```
module.exports = async function (app, opts) {
  app.get('/posts', (request, reply) => {
    return app.posts.getAll()
  })
}


// routes/v2/posts.js
module.exports = async function (app, opts) {
  app.get('/posts', async (request, reply) => {
    return {
      posts: await app.posts.getAll()
    }
  })
}
```

In the previous code, we have created an application composed of four files:

1. `server.js` starts our application.
2. `services/posts.js` creates a decorator read to all the `posts` objects from our database; note the use of the `fastify-plugin` utility to break the encapsulation.
3. `routes/v1/posts.js` implements the v1 API.
4. `routes/v2/posts.js` implements the v2 API.

There is nothing special about the prefixed routes; we can call them normally using CURL or Postman:

```
$ curl http://127.0.0.1:3000/v1/posts
[{"id":1,"title":"Hello World"}]


$ curl http://127.0.0.1:3000/v2/posts
{"posts":[{"id":1,"title":"Hello World"}]}
```

Shared business logic or code

Some routes will depend upon certain code, usually the business logic implementation or the database access. While a naïve implementation would have included this logic next to the route definitions, we moved them to a `services` folder. These will still be Fastify plugins and will use inversion-of-control mechanisms via decorators.

This approach needs to scale better as we add more complexity to our application, as we need to modify `server.js` for every single new file we add. Moreover, we duplicate the information about the prefix in two places: the `server.js` file and within the filesystem structure. The solution is to implement filesystem-based routing.

Filesystem-based routing prefixes

In order to avoid registering and requiring all the files manually, we have developed `@fastify/autoload` (`https://github.com/fastify/fastify-autoload`). This plugin will automatically load the plugins from the filesystem and apply a prefix based on the current folder name.

In the following example, we will load two directories, `services` and `routes`:

```
const fastify = require('fastify')()
fastify.register(require('@fastify/autoload'), {
  dir: `${__dirname}/services`
})
fastify.register(require('@fastify/autoload'), {
  dir: `${__dirname}/routes`
})
fastify.listen({ port: 3000 })
```

This new server.js will load all Fastify plugins in the `services` and `routes` folders, mapping our routes like so:

- `routes/v1/posts.js` will automatically have the `v1/` prefix
- `routes/v2/posts.js` will automatically have the `v2/` prefix

Structuring our code using the filesystem and services allows us to create logical blocks that can be easily refactored. In the next section, will see how we can extract a microservice from our monolith.

Splitting the monolith

In the previous section, we discovered how to structure our application using Fastify plugins, separating our code into services and routes. Now, we are moving our application to the next step: we are going to split it into multiple microservices.

Our sample application has three core files: our routes for v1 and v2, plus one external service to load our posts. Given the similarity between v1 and v2 and our service, we will merge the service with v2, building the "old" v1 on top of it.

We are going to split the monolith across the boundaries of these three components: we will create a "v2" microservice, a "v1" microservice, and a "gateway" to coordinate them.

Creating our v2 service

Usually, the simplest way to extract a microservice is to copy the code of the monolith and remove the parts that are not required. Therefore, we first structure our v2 service based on the monolith, reusing the `routes/` and `services/` folders. Then, we remove the `routes/v1/` folder and move the content of v2 inside `routes/`. Lastly, we change the port it's listening to 3002.

We can now start the server and validate that our `http://127.0.0.1:3002/posts` URL works as expected:

```
$ curl http://127.0.0.1:3002/posts
{"posts":[{"id":1,"title":"Hello World"}]}
```

It's now time to develop our v1 microservice.

Building the v1 service on top of v2

We can build the v1 service using the APIs exposed from v2. Similarly to our v2 service, we can structure our v1 service based on the monolith, using the `routes/` and `services/` folders. Then, we remove the `routes/v1/` folder and move the content of v1 inside `routes/`. Now, it's time to change our `services/posts.js` implementation to invoke our v2 service.

Our plugin uses undici (`https://undici.nodejs.org`), a new HTTP client from Node.js.

> **The story behind undici (from Matteo)**
>
> undici was born in 2016. At the time, I was consulting with a few organizations that were suffering from severe bottlenecks in performing HTTP calls in Node.js. They were considering switching the runtime to improve their throughput. I took up the challenge and created a proof-of-concept for a new HTTP client for Node.js. I was stunned by the results.
>
> How is undici fast? First, it carries out deliberate connection pooling via the use of Keep-Alive. Second, it minimizes the amount of event loop microticks needed to send a request. Finally, it does not have to conform to the same interfaces as a server.
>
> And why is it called "undici"? You could read HTTP/1.1 as 11 and undici means 11 in Italian (but more importantly, I was watching Stranger Things at the time).

We create a new `undici.Pool` object to manage the connection pool to our service. Then, we decorate our application with a new object that matches the interface needed by the other routes in our service:

```
const fp = require('fastify-plugin')
const undici = require('undici')

module.exports = fp(async function (app, opts) {
```

```
    const { v2URL } = opts
    const pool = new undici.Pool(v2URL)
    app.decorate('posts', {
      async getAll () {
        const { body } = await pool.request({
          path: '/posts',
          method: 'GET'
        })
        const data = await body.json()
        return data.posts
      }
    })

    app.addHook('onClose', async () => {
      await pool.close()
    })
  })
```

The onClose hook is used to shut down the connection pool: this allows us to make sure we have shut down all our connections before closing our server, allowing a graceful shutdown.

After creating our v2 and v1 microservices, we will now expose them via an API gateway.

Exposing our microservice via an API gateway

We have split our monolith into two microservices. However, we would still need to expose them under a single origin (in web terminology, the origin of a page is the combination of the hostname/IP and the port). How can we do that? We will cover an Nginx-based strategy as well as a Fastify-based one.

docker-compose to emulate a production environment

To demonstrate our deployment scenario, we will be using a docker-compose setup. Following the same setup as in *Chapter 10*, let's create a Dockerfile for each service (v1 and v2). The only relevant change is replacing the CMD statement at the end of the file, like so:

```
CMD ["node", "server.js"]
```

We'll also need to create the relevant package.json file for each microservice.

Once everything is set up, we should be able to build and run both v1 and v2 that we just created. To run them, we set up a `docker-compose-two-services.yml` file like the following:

```
version: "3.7"
services:
  app-v1:
    build: ./microservices/v1
    environment:
      - "V2_URL=http://app-v2:3002"
    ports:
      - "3001:3001"
  app-v2:
    build: ./microservices/v2
    ports:
      - "3002:3002"
```

After that, we can build and bring our microservices network up with a single command:

```
$ docker-compose -f docker-compose-two-services.yml up
...
```

This `docker-compose` file exposes `app-v1` on port 3001 and `app-v2` on port 3002. Note that we must set V2_URL as the environment variable of `app-v1` to tell our application where `app-v2` is located.

Then, in another terminal, we can verify that everything is working as expected:

```
$ curl localhost:3001/posts
[{"id":1,"title":"Hello World"}]
$ curl localhost:3002/posts
{"posts":[{"id":1,"title":"Hello World"}]}
```

After dockerizing the two services, we can create our gateway.

Nginx as an API gateway

Nginx is the most popular web server in the world. It's incredibly fast and reliable and leveraged by all organizations, independent of size.

We can configure Nginx as a reverse proxy for the /v1 and /v2 prefixes to our microservices, like so:

```
events {
  worker_connections 1024;
}

http {
  server {
    listen        8080;

    location /v1 {
      rewrite /v1/(.*)  /$1  break;
      proxy_pass        http://app-v1:3001;
    }

    location /v2 {
      rewrite /v2/(.*)  /$1  break;
      proxy_pass        http://app-v2:3002;
    }
  }
}
```

Let's dissect the Nginx configuration:

- The events blocks configure how many connections can be opened by a worker process.

- The http block configures our plain HTTP server.

- Inside the http->server block, we configure the port to listen to, and two locations /v1 and /v2. As you can see, we rewrite the URL to remove /v1/ and /v2/, respectively.

- Then, we use the proxy_pass directive to forward the HTTP request to the target host.

> **Nginx configuration**
>
> Configuring Nginx properly is hard. Quite a few of its settings could significantly alter the performance profile of an application. You can learn more about it from the documentation: https://nginx.org/en/docs.

After preparing the Nginx configuration, we want to start it via Docker by creating a `Dockerfile` file:

```
FROM nginx
COPY nginx.conf /etc/nginx/nginx.conf
```

Then, we can start our network of microservices by creating a `docker-compose-nginx.yml` file:

```
version: "3.7"
services:
  app-v1:
    build: ./microservices/v1
    environment:
      - "V2_URL=http://app-v2:3002"
  app-v2:
    build: ./microservices/v2
  gateway:
    build: ./nginx
    ports:
      - "8080:8080"
```

In this configuration, we define three Docker services: `app-v1`, `app-v2`, and `gateway`. We can start with the following:

```
$ docker-compose -f docker-compose-nxing.yml up
```

We can now verify that our APIs are correctly exposed at `http://127.0.0.1:8080/v1/posts` and `http://127.0.0.1:8080/v2/posts`.

Using Nginx to expose multiple services is a great strategy that we often recommend. However, it does not allow us to customize the gateway: what if we want to apply custom authorization logic? How would we transform the responses from the service?

@fastify/http-proxy as an API gateway

The Fastify ecosystem offers a way of implementing a reverse proxy with JavaScript. This is `@fastify/http-proxy`, available at `https://github.com/fastify/fastify-http-proxy`.

Here is a quick implementation of the same logic we implemented in Nginx:

```
const fastify = require('fastify')({ logger: true })

fastify.register(require('@fastify/http-proxy'), {
```

```
  prefix: '/v1',
  upstream: process.env.V1_URL || 'http://127.0.0.1:3001'
})

fastify.register(require('@fastify/http-proxy'), {
  prefix: '/v2',
  upstream: process.env.V2_URL || 'http://127.0.0.1:3002'
})

fastify.listen({ port: 3000, host: '0.0.0.0' })
```

Building an API gateway on top of Node.js and Fastify allows us to customize the logic of our gateway in JavaScript completely – this is a highly effective technique for performing centralized operations, such as authentication or authorization checks before the request reaches the microservice. Moreover, we can compile the routing table dynamically, fetching it from a database (and caching it!). This provides a clear advantage over a reverse proxy approach.

The main objection we have to building a custom API gateway with Fastify is related to security, as some companies do not trust their developers to write API gateways. In our experience, we've deployed this solution multiple times, it performed well beyond expectations, and we've had no security breaches.

After writing our proxy in Node.js, we should create the relevant Dockerfile and `package.json`. As in the previous section, we will use `docker-compose` to verify that our microservice network works appropriately. We create a `docker-compose-fhp.yml` file for this solution with the following content:

```
version: "3.7"
services:
  app-v1:
    build: ./microservices/v1
    environment:
      - "V2_URL=http://app-v2:3002"
  app-v2:
    build: ./microservices/v2
  app-gateway:
    build: ./fastify-http-proxy
    ports:
      - "3000:3000"
    environment:
```

```
    - "V1_URL=http://app-v1:3001"
    - "V2_URL=http://app-v2:3002"
```

In this configuration, we define three Docker services: `app-v1`, `app-v2`, and `app-gateway`. We can run it like so:

```
$ docker-compose -f docker-compose-fhp.yml up
```

In the next section, we will see how to customize our gateway to implement distributed logging.

Implementing distributed logging

Once we have created a distributed system, everything gets more complicated. One of the things that becomes more complicated is logging and tracing a request across multiple microservices. In *Chapter 11*, we covered distributed logging – this is a technique that allows us to trace all the log lines that are relevant to a specific request flow via the use of correlation IDs (`reqId`). In this section, we will put that into practice.

First, we modify our gateway's `server.js` file to generate a new UUID for the request chain, like so:

```
const crypto = require('crypto')
const fastify = require('fastify')({
  logger: true,
  genReqId (req) {
    const uuid = crypto.randomUUID()
    req.headers['x-request-id'] = uuid
    return uuid
  }
})
```

Note that we generate a new UUID at every request and assign it back to the `headers` object. This way, `@fastify/http-proxy` will automatically propagate it to all downhill services for us.

The next step is to modify the `server.js` file in all microservices so that they recognize the `x-request-id` header:

```
const crypto = require('crypto')
const fastify = require('fastify')({
  logger: true,
  genReqId (req) {
    return req.headers['x-request-id'] ||
    crypto.randomUUID()
```

```
    }
  })
```

The last step is making sure the invocation of the `v2` service from `v1` passes through the header (in `microservices/v1/services/posts.js`):

```
app.decorate('posts', {
  async getAll ({ reqId }) {
    const { body } = await pool.request({
      path: '/posts',
      method: 'GET',
      headers: {
        'x-request-id': reqId
      }
    })
    const data = await body.json()
    return data.posts
  }
})
```

Here, we have updated the `getAll` decorator to forward the custom `x-request-id` header to the upstream microservice.

As we have seen in this section, an API gateway built on top of Fastify allows you to easily customize how the requests are handled. While in the distributed logging case, we only added one header, this technique also allows you to rewrite the responses or add central authentication and authorization logic.

Summary

In this chapter, we discussed the problems with deploying a monolith and the different techniques to mitigate these issues: constraints and URL prefixing. The latter is the foundation for the ultimate solution: splitting the monolith into multiple microservices. Then, we showed how to apply distributed logging to a microservices world, ensuring that the requests are uniquely identifiable across different microservices.

You are ready to proceed to *Chapter 13*, in which you will learn how to optimize your Fastify application.

13

Performance
Assessment and Improvement

We all know that Fastify is fast – but is your code fast enough for Fastify? Learn how to measure your application's performance to improve your business value and the quality of your code. By analyzing the metrics, you will avoid introducing speed regressions and identify bottlenecks or memory leaks that could crash your system. You will learn how to add an instrumentation library to a Fastify application to analyze how the server reacts to a high volume of traffic. We will get an overview to understand and act accordingly depending on your measurements to maintain your server's performance and its healthy status.

This is the learning path we will take in this chapter:

- Why measure performance?
- How to measure an application's performance
- How to analyze the data
- How to optimize the application

Technical requirements

As mentioned in earlier chapters, you will need the following:

- A working Node.js 18 installation
- The VS Code IDE from `https://code.visualstudio.com/`
- A working command shell

All the snippets in this chapter are on GitHub at `https://github.com/PacktPublishing/Accelerating-Server-Side-Development-with-Fastify/tree/main/Chapter%2013`.

Why measure performance?

It sounds nice to go out for dinner and impress our guests by telling them that our APIs serve 10,000 requests per second, but nobody cares about numbers that do not bring any value.

An application's performance impacts a company's business in many ways, which are often underestimated because you can't see a performance issue at first sight as you can with a bug. In fact, performance is responsible for the following:

- A slow API, which may result in customers abandoning your website, as documented by this research: `https://www.shopify.com/enterprise/site-performance-page-speed-ecommerce`

- A server in an idle state during high-load traffic, which is a waste of resources that impacts your infrastructure bill

- Unoptimized API code, which may waste resources that affect your infrastructure bill

Therefore, to save money and open new business opportunities, it is crucial to measure the application's performance, but how can we do so? Let's start unraveling this process.

To improve an application's performance, you need to define where to start and choose which direction to follow. The **Optimization Cycle** supports you during this discovery process and focuses on the resolution:

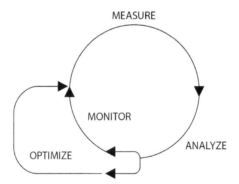

Figure 13.1 – The Optimization Cycle

Each step is a key to improving performance:

1. **Measure**: We must define a baseline to start from by selecting certain **Key Performance Indicators** (**KPIs**) that can be measured and evaluated further.

2. **Analyze**: We need to understand and explore the KPIs collected in the previous step to define how to improve them and find any bottlenecks. When the KPI satisfies our expectations, we can just monitor the performance without taking any other action.

3. **Optimize**: We apply the required actions identified in the previous step. We may make changes to our code base or even to the application's architecture. In the end, we will start again at the first step to evaluate the results.

By following this cycle, you will be able to assess the application's performance and keep it under control in order to avoid regression. Note that this workflow adapts to every area of a company's work, from sales to information technology.

We have seen the overall picture of this methodology, but what happens when it's in action? We are going to discover it in the next section.

How to measure an application's performance?

By following the Optimization Cycle, the first step to improving something in every work field is that we need to set a baseline. The starting point will be different for each company's department. For instance, the sales team will measure the sales or the revenue over time to set its improvement goals.

So, we need to define what is worth measuring by defining our application's metrics boundaries. It's clear that our first challenge is choosing what to measure. The most common metrics are the following:

- HTTP response times and request rates
- Error ratings
- Process metrics such as CPU, memory, and Garbage Collector usage

Each bullet point is a KPI. Every KPI requires a different approach and a different analysis. Despite their differences, these measurements are strongly connected since inadequate resource management will impact the response latency, and a high error rating may increase memory usage.

Measuring an application's KPIs is a mandatory step in implementing application **Observability**, which is the capability to understand the system's state by checking its output.

> **Application Observability**
>
> The following section's concepts apply to a whole system or a single Node.js application. This chapter discusses Application Observability and aims at improving a project itself, as system monitoring is not in the scope of this book. Nevertheless, what you will learn will be valuable and reusable for building better systems.

The pillars of Observability are as follows:

- **Logs**: You can learn about an application's status and progress by reading and analyzing the output. We discussed this in *Chapter 11*.

- **Metrics**: These are numeric value representations that an application can collect. These are more reliable, ready to be processed, and easier to store. Compared to logs, they have an indefinite retention policy. You collected them in *Chapter 10*.

- **Tracing**: This represents a piece of detailed information from the request life cycle over the system. We will see an example in the next section, *Measuring the HTTP response time by instrumenting the Fastify application*.

All these pillars share a common backbone of logic. We can consider all of them raw logs in different formats and properties that share the same life cycle logic as we saw in the *How to consolidate the logs* section of *Chapter 11*.

The pillar list is ordered from the least detailed to the most fine-grained information. I think an example is worth more than 10,000 words, so let's assume we have to assess the POST /todo endpoint that we have been building since *Chapter 7*.

In this case, the logs would tell us when the request started and when it ended by adding debugging information.

A metric tells us what the KPI value at a specific time was. For example, it answers the question, how many todos were created per user today? This information is easy to process if you want to know the average number of *todos* created and other statistics.

The tracing data would show us that a request spent 10 milliseconds fetching the data from the database, 20 milliseconds inserting the new data, and 30 additional milliseconds sending a notification.

To understand this example and appreciate the differences better, let us focus on how to measure the HTTP response time in the next section.

Measuring the HTTP response time

The KPI of response time is a simple parameter that even non-technical people can easily understand, and it is commonly used to describe an API's health status. It reports the total amount of time the application spent handling an HTTP request and replying to the client.

The response time provides us with a precise measure in milliseconds to monitor the overtime for each endpoint. By noticing any variation in this value, we can evaluate whether a new feature impacts the route's health positively or negatively. We can then act accordingly, as we will see in the *How to analyze the data* section.

Network latencies

The API response time can sometimes be misleading because it does not include additional network latencies that are introduced if your server is in a different region from the client. Discussing complex architectural scenarios is not within the scope of this book. If you want to go deeper into this aspect, you can start by reading this article: `https://www.datadoghq.com/knowledge-center/infrastructure-monitoring/`.

We are already measuring the overall request time by logging the `responseTime` field, as explained in *Chapter 11*. This information can be read directly from the production environment and can be checked constantly. Moreover, it helps us set which APIs have priority over others, as they need more attention. This aspect is beneficial considering that we can't focus on hundreds of endpoints at the same time, so this information will help us prioritize certain APIs.

The response time is not detailed information because it is the sum of all the steps performed by our endpoint to reply to the client's request. For example, if our API makes a database query and an external HTTP call to a third-party API, which of these two operations costs more time?

Since we are arranging the data to analyze it in the next Optimization Cycle step, we must collect more information by **tracing** the application; otherwise, the analysis task will be more complex and optimization may not be successful.

> Tracing sampling
>
> While logs have a resource cost, tracing is even more expensive because it produces huge data volumes. For this reason, when dealing with tracing, it is mandatory to set the sampling configuration. You will see an example in this section.

Now that we know what we want to measure, we can start to upgrade our Fastify application.

Instrumenting the Fastify application

To trace any Fastify application, it is wise to adopt the **OpenTelemetry** specification published at `https://opentelemetry.io/` and promoted by the Linux Foundation. This provides a set of shared mechanisms for generating, collecting, and exporting telemetry data, such as metrics and tracing information. Even though logs are currently not supported by this tool, we are safe because we have already set the application's logs in *Chapter 11*.

By adopting the OpenTelemetry specification, you will be able to include modules that implement this standard, and your application will export metrics and tracing data in a well-known format. By choosing this approach, you decouple your telemetry data from the **Application Performance Monitoring** (**APM**) tool. The APM software is a must-have for analyzing your tracing data. It provides a straightforward way to see the telemetry data and put it into graphs and clear visualizations. Otherwise, it would be impossible to understand the raw tracing data and the connections within it.

It is now time to write some code! To recap what we are going to do, here are the steps:

1. Integrate the OpenTelemetry modules into your Fastify to-do list application.

2. Visualize the telemetry data by using Zipkin, an open source tracing system. It is less powerful than a commercial APM, but it is a free solution for every developer starting this journey.

3. Improve the tracing and metric information.

Before we start, it is mandatory to install all the OpenTelemetry packages we need, so you need to run the following installation command:

```
npm install @opentelemetry/api@1.3.0 @opentelemetry/exporter-
zipkin@1.8.0 @opentelemetry/instrumentation@0.34.0 @opentelemetry/
instrumentation-dns@0.31.0 @opentelemetry/instrumentation-
fastify@0.31.0 @opentelemetry/instrumentation-http@0.34.0 @
opentelemetry/instrumentation-mongodb@0.33.0 @opentelemetry/sdk-
node@0.34.0 @opentelemetry/sdk-trace-node@1.8.0 @opentelemetry/
semantic-conventions@1.8.0
```

These are a lot of modules, but we are going to use them all in the upcoming script. We need to create a new `configs/tracing.js` file that will contain all the tracing logic.

The script will be split into three logic blocks. The first one is dedicated to imports:

```
const packageJson = require('../package.json')
// [1]
const { NodeTracerProvider } =
  require('@opentelemetry/sdk-trace-node')
const { SemanticResourceAttributes } =
  require('@opentelemetry/semantic-conventions')
const { Resource } = require('@opentelemetry/resources')
const { ParentBasedSampler, TraceIdRatioBasedSampler } =
  require('@opentelemetry/sdk-trace-base')
  // [2]
const { registerInstrumentations } =
  require('@opentelemetry/instrumentation')
const { DnsInstrumentation } =
  require('@opentelemetry/instrumentation-dns')
const { HttpInstrumentation } =
  require('@opentelemetry/instrumentation-http')
const { FastifyInstrumentation } =
  require('@opentelemetry/instrumentation-fastify')
const { MongoDBInstrumentation } =
  require('@opentelemetry/instrumentation-mongodb')
  // [3]
const { BatchSpanProcessor } =
  require('@opentelemetry/sdk-trace-base')
const { ZipkinExporter } =
  require('@opentelemetry/exporter-zipkin')
```

The `require` statements are mainly the modules, [1], which provide the OpenTracing APIs to set up the system. The requirements, [2], are the `Instrumentation` classes, which will help us trace some packages for use, saving us time and providing a great starting point. We are going to see the output in a while. Finally, the `Exporter` component, [3], is responsible for submitting the tracing data to any system.

The second code block logic is the OpenTelemetry configuration:

```
const sdk = new NodeTracerProvider({
  sampler: new ParentBasedSampler({
    root: new TraceIdRatioBasedSampler(1)
  }),
  resource: new Resource({
    [SemanticResourceAttributes.DEPLOYMENT_ENVIRONMENT]:
    process.env.NODE_ENV,
    [SemanticResourceAttributes.SERVICE_NAME]:
    packageJson.name,
    [SemanticResourceAttributes.SERVICE_VERSION]:
    packageJson.version
  })
})
registerInstrumentations({
  tracerProvider: sdk,
  instrumentations: [
    new DnsInstrumentation(),
    new HttpInstrumentation(),
    new FastifyInstrumentation(),
    new MongoDBInstrumentation()
  ]
})
```

The configuration has the `resource` parameter to configure some basic metadata that will be displayed for every trace. The `instrumentations` array injects the code into our application's core module, such as Node.js's `http` package or the `fastify` installation to monitor our application.

Generally, these classes perform **monkey patching** – this technique modifies the internal Node.js cache and acts as a man-in-the-middle. Although some versions of Node.js provide experimental `async_hook` and `trace_events` modules, monkey patching is still broadly used by tracing packages. The `sampler` option is a controller to filter out certain tracing events to reduce the amount of stored data. The example configuration is a pass-all filter that you need to customize after becoming more confident using this tool.

The last code block to complete the `tracing.js` file is for the usage of the `sdk` variable:

```
const exporter = new ZipkinExporter({
  url: 'http://localhost:9411/api/v2/spans'
})
sdk.addSpanProcessor(new BatchSpanProcessor(exporter))
sdk.register({})
console.log('OpenTelemetry SDK started')
```

At first, `exporter` will transmit or expose the data to any external system. Note that if `exporter` submits the data, it will implement Push logic. Instead, if `exporter` exposes an HTTP route (which the external system must call to get the data), it will implement Pull logic.

The `register` function applies monkey patching to our application's modules to start getting all the traces. Its architecture, the OpenTelemetry script, must be run before our application loads, so we need to edit `package.json`'s `start` script by adding a new argument:

```
"start": "fastify start --require ./configs/tracing.js -l info
--options app.js",
```

By doing so, the OpenTelemetry script will run before the Fastify application. For this reason, the `@fastify/env` plugin will not start right away and it may be necessary for us to load the `.env` file into our `tracing.js` file.

Node.js standard arguments

The Fastify CLI emulates the Node.js CLI's `--require` argument. For this reason, if you are not using the Fastify CLI, you can use the same option to integrate OpenTelemetry within your application. More detail can be found on the official documentation at `https://nodejs.org/api/cli.html#-r---require-module`.

At this stage, we can still run our application as usual with `npm start`, but without storing any tracing data because `exporter` is not configured yet. Let's see how to do that.

Visualizing the tracing data

We are quite ready to start the tracing, but we need a Zipkin instance on our system. We can use `docker` for this task. Add these new utility commands to `package.json`:

```
    "zipkin:start": "docker run --rm --name fastify-zipkin -d -p
9411:9411 openzipkin/zipkin:2",
    "zipkin:stop": "docker container stop fastify-zipkin",
```

After that, run `npm run zipkin:start` in your shell and you should be able to reach the `http://localhost:9411/zipkin` URL from your browser.

After running the npm run mongo:start && npm start command, the application is ready to call the GET /todos endpoint, and by pressing Zipkin's **RUN QUERY** button, you should see this output:

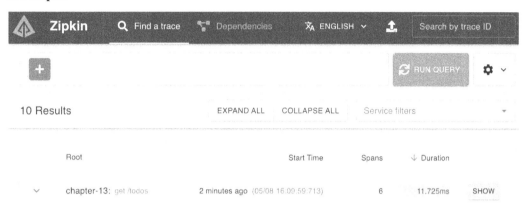

Figure 13.2 – The Zipkin home page

From Zipkin's home page, we have an overview of all the trace logs we are capturing from our application. Now, if you press the **SHOW** button for the get /todos endpoint, you will be able to see the details:

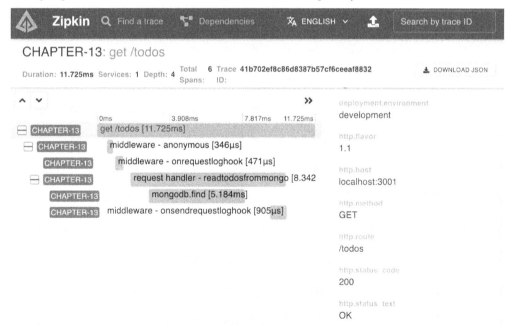

Figure 13.3 – The request detail page

From the detail page, you have a complete overview of the hooks that have been executed, and you can evaluate the timing of every step of your handler.

> **The importance of named functions**
>
> In *Figure 13.3*, you can see an anonymous middleware function: a hook within an arrow function. Remember that when it's not strictly necessary, you can adopt a named function to read the functions' names in your APM software, and this will make debugging much smoother. We can also see `onrequestloghook` in *Figure 13.3*, which we defined in *Chapter 11*.

Now, we can use the Zipkin UI as a starting point to search for endpoints that require a follow-up analysis and may need to be optimized. Many commercial APM tools offer features that monitor and send you an alert directly, without the need to dig into raw data. In this section, you have covered the core concepts that will support you in choosing a solution that fits your needs. By adopting the OpenTelemetry specification, your application will be vendor-agnostic.

Congratulations – now, we can measure every route's execution detail. We can proceed to the next phase in the Optimization Cycle: the analysis.

How to analyze the data

Following the Optimization Cycle, we must analyze the measurement result to understand the following:

- Can the application performance be improved?
- Where should we focus first?
- How can we improve it?
- Where is the bottleneck?

For example, if we analyze *Figure 13.3*, the handler spends most of its time on the MongoDB `find` operation. This means that we should focus on that part, trying to check whether the following applies:

- We have set the connection pool correctly
- We have created a dedicated index

Based on whether this is the case or not, you can define some boundaries that can be further optimized in the next Optimization Cycle step.

Of course, when monitoring all the application's routes, we will need many filters to select all the routes that require our attention, such as slow endpoints or APIs with a high error rate.

Note that, at times, the data cannot be as detailed as in this example. Let's assume that the measurement says that one specific endpoint performs terribly. In this case, we must analyze the target handler in detail.

To analyze it in detail, certain tools can help us with this difficult task. Let's start by defining our local baseline using autocannon (https://www.npmjs.com/package/autocannon). This module is a benchmarking tool to stress test an endpoint. Let's see the output by running the commands:

```
npm install autocannon@7 -g
npm pkg set scripts.performance:assessment="autocannon -d 20 -c 100"
npm run performance:assessment -- http://localhost:3000/todos
```

This performance assessment will run 100 connections within 20 seconds and will produce an excellent report as follows:

```
> chapter-13@1.0.0 performance:assessment
> autocannon -d 20 -c 100 "http://localhost:3001/todos"

Running 20s test @ http://localhost:3001/todos
100 connections
```

Stat	2.5%	50%	97.5%	99%	Avg	Stdev	Max
Latency	5 ms	20 ms	52 ms	72 ms	21.54 ms	13 ms	234 ms

Stat	1%	2.5%	50%	97.5%	Avg	Stdev	Min
Req/Sec	2993	2993	4607	5047	4536.9	480.75	2992
Bytes/Sec	3.37 MB	3.37 MB	5.19 MB	5.69 MB	5.11 MB	542 kB	3.37 MB

```
Req/Bytes counts sampled once per second.
# of samples: 20

91k requests in 20.02s, 102 MB read
```

Figure 13.4 – An autocannon report

At this point, we have a baseline: the GET /todos endpoint in *our working machine* manages approximately 4,500 requests per second on average and has a 20-millisecond latency. Of course, there is no absolute value to aim for; it depends entirely on the API's business logic and its consumers. As a general rule, we can say the following:

- Up to approximately 150 milliseconds of latency, the API is responsive
- Up to 300 milliseconds, the API is acceptable
- If it is 1,000 milliseconds or above, the API may impact your business negatively

Now, we can search for more details by installing a new tool: Clinic.js (`https://clinicjs.org/`). The following commands are run to install the module and store some new `package.json` scripts:

```
npm install clinic@12 -D
npm pkg set scripts.clinic:doctor="clinic doctor --on-port 'autocannon
-d 20 -c 100 http://localhost:3000/todos' -- node index.js"
npm run clinic:doctor
```

The `clinic:doctor` command will do the following:

1. Start the application.

2. Run the `autocannon` test.

3. Build and open an excellent HTML report automatically when you stop the application manually.

You may have noticed the `node index.js` usage in the `clinic:doctor` script. Unfortunately, Clinic.js does not support any command other than `node`. For this reason, it is mandatory to create a new `index.js` file on the project's root directory, as follows:

```
const { listen } = require('fastify-cli/helper')
const argv = ['-l', 'info', '--options', 'app.js']
listen(argv)
```

The previous script skips the `tracing.js` initialization because the focus of this analysis phase is on a single endpoint or aspect of an application and the tracing instrumentation could make our deep analysis harder, as it adds noise to the output shown in *Figure 13.5*.

Upon the completion of the `clinic:doctor` command, you should see something similar to this graph in your browser:

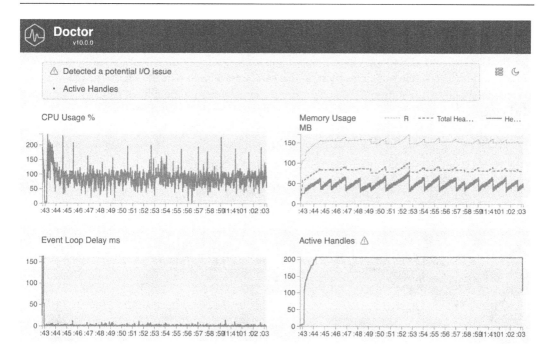

Figure 13.5 – Output of the Clinic.js report

As you can see, Clinic.js is very detailed and provides us with an overview of the entire performance. It is so smart that it even suggests to us where to focus our attention and why:

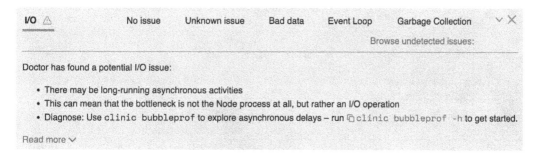

Figure 13.6 – Clinic.js suggestions

By reading the Clinic.js output, we can get an overview of our application while autocannon stress tests it. We can go deeper into each graph from this big picture with Clinic.js commands in the next sections.

Creating a flame graph

When the CPU and the event loop usage are too high, there can be issues with our code base or the external modules installed in our application. In this case, we will need to make a **flame graph**, which shows us the functions' execution time as percentages. By abstracting the time, it makes it easier to find a bottleneck.

As an example, we can run this command:

```
clinic flame --on-port 'autocannon -d 20 -c 100 http://localhost:3000/
todos' -- node index.js
```

When it has run, you will see something like the following:

Figure 13.7 – A flame graph

The graph in the preceding figure shows a healthy application. The *x*-axis represents the entire execution time, every stacked bar is the function call stack, and the length of a bar represents the time spent by the CPU on each function. This means that the wider the bar, the more the function slows down our application, so we should aim for thick bars.

As an experiment, we can modify the GET /todos handler to slow the handler function and look at a bad flame graph as a comparison, so let's add this code line to the routes/routes.js file:

```
fastify.get('/todos', {
  handler: async function readTodosFromMongo (request,
  reply) {
    require('fs').readFileSync(__filename) // waste CPU
                                                  cycles
    // ...
  }
```

After this code change, we can rerun the clinic flame command and wait for the report:

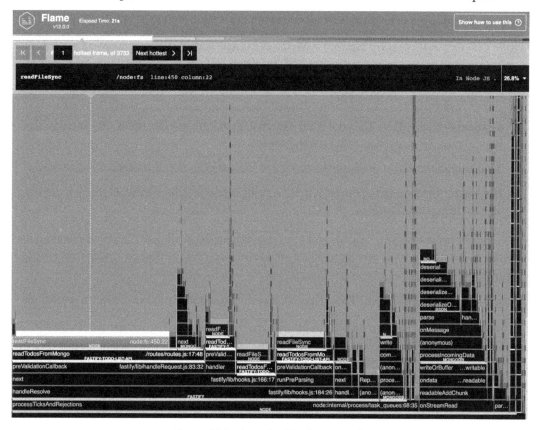

Figure 13.8 – An unhealthy flame graph

As you can see in *Figure 13.8*, a significant horizontal bar occupies about 27% of the CPU's time! This would be a function that needs to be optimized. By reading the function's call stack, we would find the issue's source immediately – the routes/routes.js file – and fix the code.

A flame graph is a powerful tool for finding issues related to our code base and solving high CPU usage and high event loop delay.

We are not done with the Clinic.js commands yet – let's dive into the next section to discover the next one.

How to check memory issues

The `clinic doctor` report in *Figure 13.5* shows the memory usage over time. It helps us understand whether our application has a **memory leak** that could lead to memory exhaustion or a crash of the process. A memory leak happens when the program allocates the system's memory by defining variables or closures and, after its usage, the Node.js garbage collector cannot release this allocated memory. Usually, this situation happens when the code does not release the instantiated objects.

In *Figure 13.5*, we can see positive memory behavior, as its trend is stable because the mean value does not grow over time, as you can see in *Figure 13.10*. The drops you see in the same figure are triggered by Node.js's garbage collector, which frees up the unused object's memory. In the same image, in the upper-right graph, we can see three lines:

- **Resident Set Size (RSS)**: The total RAM allocated for the process execution. This includes the source code, the Node.js core objects, and the call stack.

- **Total heap allocated**: The memory allocated that we can use to instantiate the application's resources.

- **Heap used**: The total size of the heap space actually in use, which contains all the application's objects, strings, and closures.

I will not bother you with how Node.js and the V8 engine underneath work, but you may find it interesting to go into this topic deeper by reading this article: `https://deepu.tech/memory-management-in-v8/`.

Clinic.js gives us another tool to check the application's memory usage. Try to run the `heap` command:

```
clinic heap --on-port 'autocannon -d 20 -c 100 http://localhost:3000/
todos' -- node index.js
```

As usual, we get back a nice report to read as follows:

Figure 13.9 – A heap graph

The application's memory usage in *Figure 13.9* can be read as a flame graph, as discussed in the *Creating a flame graph* section. The *x*-axis represents the total amount of our application execution, and the bigger the function's bar is, the more memory is allocated to it. In this case, we must also check the biggest bar to reduce memory usage.

For example, let's try and add a memory leak into the GET /todos route:

```
const simpleCache = new Map()
fastify.get('/todos', {
  handler: async function readTodosFromMongo (request,
  reply) {
    const cacheKey = 'todos-' + request.id
    if (simpleCache.has(cacheKey)) {
      return simpleCache.get(cacheKey)
    }
    const todos = Array.from(1e6).fill('*')
    simpleCache.set(cacheKey, todos)
    return todos
```

```
    }
  })
```

In the previous, wrong code example, we wanted to add a cache layer to reduce the number of queries we were executing. Unfortunately, we created the `cacheKey` with a typo. For this reason, the `simpleCache.has()` check will always be false, and the `simpleCache` object will continue to add new keys. Moreover, we did not implement a cache clearing method, so the `simpleCache` object will continue to grow. By running the `clinic` command first, we will get a clear memory leak graph:

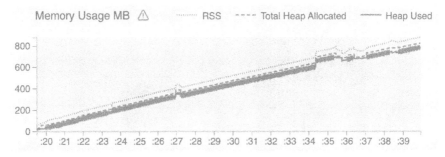

Figure 13.10 – Memory leak overview

However, if we run the `clinic heap` command instead, we will get more precise detail about the issue we have introduced:

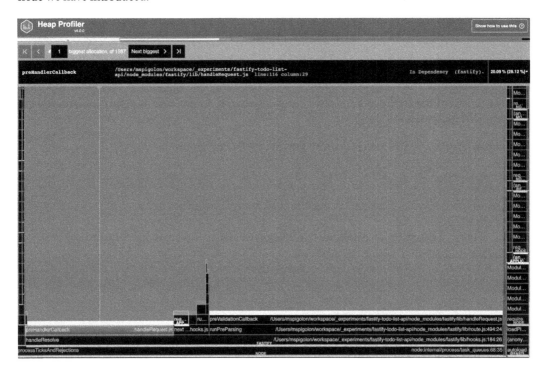

Figure 13.11 – Memory leak details

Looking at the heap graph, we can see three big horizontal bars caused by our incorrect implementation of caching. The `heap` command is crucial to solving memory issues, as its capability points us in the right direction. Clinic.js provides us with a fantastic tool, but the surprises are not over yet. Let's see another useful tool.

How to identify I/O resources

Figure 13.5 shows us an **Active Handles** graph, which counts the file descriptors opened as files and sockets. It warns us to keep our I/O resources under control. In fact, we did an `autocannon` test and set 100 concurrent connections, so why is that graph showing us 200 active handlers?

To figure this out, we need to use the `clinic bubble` command:

```
clinic bubble --on-port 'autocannon -d 20 -c 100 http://
localhost:3000/todos' -- node index.js
```

The bubble report is a new kind of report that is different from the previous graphs we built. Here is the output:

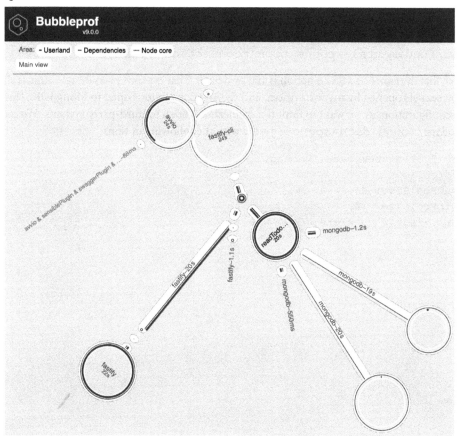

Figure 13.12 – A bubble report

Each bubble represents I/O time, so the bigger the bubble, the more time you have to spend on it. This graph looks quite good actually. There is no obvious bigger bubble to analyze but if you look closer, you will see that the green bubbles are both related to MongoDB, and one of them is the biggest one.

> **Troubleshooting**
>
> You will need to turn off Fastify's logger to generate the bubble graphs. Clinic.js does not support Node.js worker threads, and the `pino` logger module uses those features. This incompatibility would generate an unreadable graph. Moreover, you may encounter the `Analysing dataError: premature close` error during report generation. This is a known problem based on the operating system and Node.js version, as described here: `https://github.com/clinicjs/node-clinic-bubbleprof/issues/399`. As a solution, you can manually apply the quick fix.

By clicking on the bubble, we can explore the graph and spot that the `"makeConnection@mongodb"` async operation is the slowest. Here, we don't have a clear solution as we did with the previous heap and flame graphs, but we need to think about the hints that this bubble graph is giving us.

Looking closer at the MongoDB configuration, we did not set any options but read the default client's values from the official website at `https://www.mongodb.com/docs/drivers/node/current/fundamentals/connection/#connection-options`.

We can see that the `maxPoolSize` option is 100 by default. That explains everything! Clinic.js shows us the 100 sockets opened by `autocannon`, and the 100 connections open to MongoDB! This is a clear misconfiguration, as we want to limit the connections open to third-party systems. We can fix the `plugins/mongo-data-source.js` file with the following options:

```
fastify.register(fastifyMongo, {
  forceClose: true,
  maxPoolSize: 20,
  minPoolSize: 10,
  url: fastify.secrets.MONGO_URL
})
```

Now, if we rerun the `bubble` command, we can see that a big bubble has disappeared:

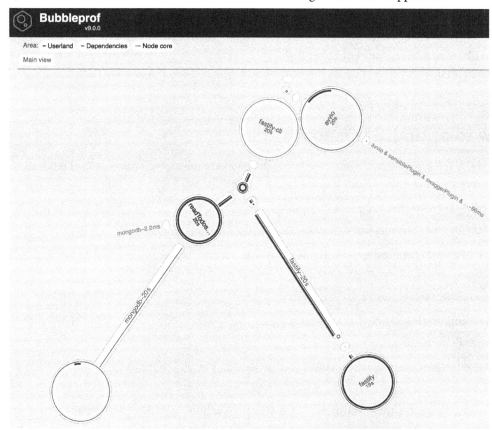

Figure 13.13 – The MongoDB connection issue solved

The new bubble graph looks even better than the one in *Figure 13.12*, and by checking a new doctor run, we can say that the **Active Handles** counter is as expected and is under control:

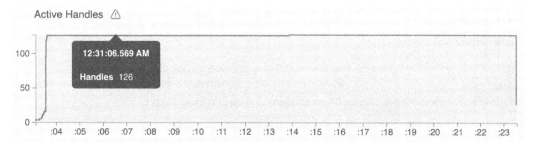

Figure 13.14 – Active handlers after the MongoDB max pool configuration

The preceding figure shows us the 100 connections and the maximum of 20 connections to MongoDB. This optimization will protect our database from potential attacks that aim to saturate the system's network and reuse the connection from the pool, saving time.

The `bubble` command is the last secret that Clinic.js unveiled to us. From now on, I hope you will have a new point of view on your application configuration and all of its moving parts.

In the next section, we are going to explore the last step of the Optimization Cycle.

How to optimize the application

To optimize the application, we must start from the output of the analysis phase of the Optimization Cycle workflow. The actions that we could take are very different:

- **No action**: The data analysis suggests that optimization is not possible or is not worth the effort to improve performance
- **Code base optimization**: The data tells us that there are bad algorithms or bad business logic implementation in our code base, and we can rewrite it to speed up the server
- **Architectural action**: The data identifies a system bottleneck, such as a legacy database
- **Configuration setup**: The analysis suggests that tweaking the configuration could improve the performance

The **No action** step is continually revised based on data that we will continue to measure and analyze over time.

For every action type, there is different knowledge to acquire. Working within a fantastic and heterogeneous team helps cover all these aspects, which are not always related to a developer's daily job. In the *How to analyze the data* section, we already saw an example for each use case:

- **Code base optimization**: This is solved using a flame graph
- **Architectural action**: This has been covered by introducing a cache to reduce the number of database queries
- **Configuration setup**: The bubble graph helps us find the proper configuration to apply to MongoDB

For this reason, we will focus on code base optimization by mentioning all the best practices and the common issues I have found in my experience of many, many hours spent analyzing metrics data. The following sections collect a series of suggestions and insights you need to know.

Node.js internals that you need to know

The JavaScript object model is based on the internal concepts of `Shape` constructs. Every object has at least one `Shape` construct that maps the object's properties to the memory offset where the value is stored. Objects can share the same `Shape`, leading to better performance by reducing memory usage and memory access.

Problems occur when an object seems to have a single `Shape`, but multiple `Shape` constructs are created due to some trivial coding error. Therefore, every time you mutate `Shape` by adding or removing properties, you impact the overall performance of the application.

Therefore, if the analysis has found some memory issues, such as frequent garbage collection, you should check how objects are built and processed in the application.

As a general rule, you should declare all the object's properties during the declaration. The following is unoptimized code:

```
const object = {}
if (something()) {
  object.value = 1
} else {
  object.key = 'value'
  object.value = 2
}
```

To optimize the previous code snippet, it is enough to write: `const object = { value: undefined, key: undefined }`. Similarly, if you want to remove a property, you can set it to `undefined` instead.

To strengthen these Node.js core concepts, you may want to refer to this great video – `https://mathiasbynens.be/notes/shapes-ics` – made by Mathias Bynens, who is a Google developer and TC39 member.

The safety of parallel running

Sometimes, it is easier to think of a sequence of steps that our software will take, but this is not the case. A typical example is the following one:

```
const user = await database.readUser(10)
const userTasks = await database.readUserTask(user)
const userNeeds = await database.readUserNeeds(user)
```

This can be optimized by using the `Promise.all` method:

```
const [userTasks, userNeeds] = await Promise.all([
  database.readUserTask(user),
```

```
    database.readUserNeeds(user)
  ])
```

If you have an array of users, you could do the same, but you must evaluate the array length. So, if your array has more than 100 items, you cannot create 1,000 promises, because this may cause you to hit a memory limit error, especially if the endpoint is called multiple times concurrently. Therefore, you can split your array into smaller ones and process them one group at a time. For this task, the p-map module (`https://www.npmjs.com/package/p-map`) works great.

Connection management takeaways

When we work with an external system, we must consider that establishing a connection is heavy. For this reason, we must always be aware of how many connections our application has to third-party software and configure it accordingly. My checklist looks like this:

- Does the external system Node.js module support a connection pool?
- How many concurrent users should every application instance serve?
- What are the external system connection limits?
- Is it possible to configure the connection name?
- How is the application server's network configured?

These simple questions helped me lots of times defining the best network settings, mainly databases.

> **Beware of premature optimization**
>
> You may fall into the premature optimization trap now that you know about code optimization. This trap makes your work harder and more stressful if you try to write optimized software. Therefore, before writing code that runs parallel or caches a lot of data, it is important to start simply by writing a straightforward handler function that just works.

Now, you are aware of all the aspects to think of when you need to optimize your application's performance.

Summary

In this dense chapter, we touched on many concepts that will help us measure, analyze, and optimize an application's performance. This Optimization Cycle is a practical workflow to follow step by step for this challenging task, which may lead to excellent business achievements, such as improving sales or overall customer satisfaction.

Now, you can evaluate a tracing module. By knowing how it works under the hood, you can integrate it into any application to start the measurement process. You also know which third-party software you will need to process and visualize the data, such as an APM. Moreover, you are now able to dig deeper into any performance issue, by analyzing it in detail, supported by some open source tools. Last but not least, once you have found a problem, you can plan an optimization to solve the issue and boost your application's endpoints iteratively.

You have completed this journey toward building a performant and maintainable Fastify application! It is time for you to push some code into production!

Up next is the last part of this book, which will discuss some advanced topics. The first one will be GraphQL. In the next chapter, you will learn about GQL and how to run it on top of your Fastify application.

Happy coding with Fastify!

Developing a GraphQL API

GraphQL is growing in popularity, and every day, more and more services expose their API using this query language. The GQL API interface will help your API consumers to retrieve the minimal set of data they need, benefiting from intuitive and always up-to-date documentation. GraphQL is a first-class citizen in the Fastify ecosystem. Let's learn how to add GraphQL handlers using a dedicated plugin, avoiding common pitfalls and taking advantage of Fastify's peculiar architecture.

Here's the learning path we will cover in this chapter:

- What is GraphQL?
- Writing the GQL schema
- How to make live a GQL schema?
- How to improve resolver performance?
- Managing GQL errors

Technical requirements

To complete this chapter successfully, you will need:

- A working Node.js 18 installation
- The VS Code IDE from `https://code.visualstudio.com/`
- A working command shell

All the snippets in this chapter are on GitHub at `https://github.com/PacktPublishing/Accelerating-Server-Side-Development-with-Fastify/tree/main/Chapter%2014`.

What is GraphQL?

GraphQL is a new language that has changed how a web server exposes data and how the client consumes it. Considering our application's data structure, we could map every data source to a graph of nodes (objects) and edges (relations) to connect them.

Here's a quick example of a GraphQL query that maps a family and its members:

```
query {
  family(id: 5) {
    id
    members {
      fullName
      friends {
        fullName
      }
    }
  }
}
```

By reading our first GraphQL query, we can immediately understand the relation hierarchy. A `Family` entity has many `Person` as a `members` array property. Each item of `members` may have some `Person` entities as `friends`. Commonly, a GQL request string is called a **GQL document**.

The JSON response to our GQL query request could be as follows:

```
{
  "data": {
    "family": {
      "id": 5,
      "members": [
        {
          "fullName": "Foo Bar",
          "friends": []
        },
        {
          "fullName": "John Doe",
          "friends": [
            { "fullName": "Michael Gray" },
            { "fullName": "Greta Gray" }
          ]
        },
        {
          "fullName": "Jane Doe",
          "friends": [
            { "fullName": "Greta Gray" }
```

```
            ]
          }
        ]
      }
    }
  }
}
```

Seeing the previous quick example, you may guess that if you wanted to get the same data by using a REST API architecture, you should have executed many HTTP calls, such as:

- Calling the GET /family/5 endpoint to get the family members

- Calling GET /person/id for each member to get the person's friends

This easy communication is guaranteed because GraphQL is a declarative, intuitive, and flexible language that lets us focus on the data shape. Its scope is to develop a structured and productive environment to simplify the client's API fetching.

To reach its goals, it has many design principles:

- **Product-centric**: The language is built around the consumers' requirements and data visualization

- **Hierarchical**: The request has a hierarchical shape that defines the response data structure

- **Strong-typing**: The server defines the application type system used to validate every request and document the response results

- **Client-specified response**: The client knows the server capabilities and which one it is allowed to consume

- **Introspective**: The GraphQL service's type system can be queried using GraphQL itself to create powerful tools

These principles drive the GraphQL specification, and you can find more about them at http://spec.graphql.org/.

But what do you need to implement the GraphQL specification? Let's find out in the next section.

How does Fastify support GraphQL?

GraphQL describes the language, but we must implement the specification to support its grammar. So, we are going to see how to implement GraphQL in Fastify while we explore the specification. We will write the source code to support the GQL example we saw in the previous section.

First, we need to identify the components. The following diagram shows the architectural concepts that support GraphQL:

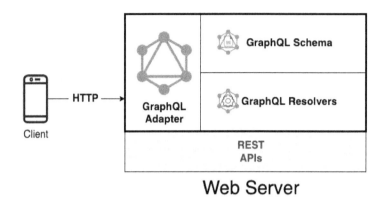

Figure 14.1 – Basic GraphQL component architecture

Figure 14.1 shows us a few essential concepts about GraphQL:

- Any client can execute a GraphQL document by performing an HTTP request to the web server.

- The web server understands the GQL request, using a **GraphQL adapter** that interfaces with the **GraphQL schema** definition and the **GraphQL resolvers**. You are going to learn all these concepts further in the *How to make live a GQL schema?* section.

- A web server could expose some REST APIs besides the GQL one without any issues.

The straightforward process we are going to follow to implement our first GraphQL server is as follows:

1. Define the GQL schema.

2. Write a simple Fastify server.

3. Add the `mercurius` plugin, and the GraphQL adapter designed for Fastify, to the Fastify installation.

4. Implement the GQL resolvers.

So, let's start working on the GQL syntax to write our first schema.

Writing the GQL schema

We must write the GQL schema by using its **type system**. If we think first about our data, it will be easier to implement it. Let's try to convert the following diagram to a schema:

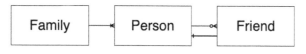

Figure 14.2 – Data relation

The data relation in *Figure 14.2* describes the entities and the relations between them:

- A *family* has multiple members
- A *person* may have numerous *friends* that are other *family* members

So, we can represent these entities into **object types** by writing the following **Schema Definition Language (SDL)**:

```
type Family {
  id: ID!
  name: String!
  members: [Person!]!
}
type Person {
  id: ID!
  family: Family!
  fullName: String!
  nickName: String
  friends: [Person!]!
}
```

The syntax to define a GQL entity is easily readable. We have defined two types. Each type has a `PascalCase` name and a list of **fields** surrounded by curly braces.

Each field is defined by `<field name>: <field type>`. The field type defines the response field's value type. You may have seen a trailing exclamation mark in the preceding code block. It is a **GQL type modifier** that declares the field as not nullable. The possible syntaxes are:

- `<field name>: <field type>!`: Not nullable field. The client must always expect a response value.
- `<field name>: [<field type>]`: The field returns a nullable array with nullable items.
- `<field name>: [<field type>!]`: The returned array is nullable, but it will not have any nullable items.
- `<field name>: [<field type>!]!`: Defines a not nullable array without any `null` items. It can still be an empty array as a result.

The field type can be of another type defined by ourselves, as we did for the `family` one, or it can be a **scalar**. A scalar represents a primitive value, and by default, every GQL adapter implements:

- `Int`: Represents a signed 32-bit numeric.

- `Float`: Represents floating-point values.

- `String`: Text data value represented as UTF-8 chars.

- `Boolean`: True or false value.

- `ID`: Represents a **unique identifier** (**UID**). It accepts numeric input, but it is always serialized as a string.

The specification lets you define additional scalars, such as `Date` or `DateTime`.

We have now written our GQL object types, but how can we use them to read and edit them? Let's find out in the next section.

Defining GQL operations

A GQL document may contain different **operations**:

- `query`: A read-only action

- `mutation`: A write and read operation

- `subscription`: A persistent request connection that fetches data in response to events

These operations must be defined in the GQL schema to be consumed by the client. Let's improve our SDL by adding the following code:

```
type Query {
  family(id: ID!): Family
}
type Mutation {
  changeNickName(personId: ID!, nickName: String!): Person
}
type Subscription {
  personChanged: Person
}
```

The previous code snippet adds one operation for each type, supported by the GQL specification. As you can read, we used the `type <operation>` syntax. These types are called **root operation types**. Each field in these special types' definitions will match a resolver that implements our business logic.

In addition to what we learned in the *Writing the GQL schema* introduction section about the `type` definition, we can see that some fields have input parameters: `family(id: ID!): Family`. In fact, the syntax is the same as we discussed previously, but there is one additional argument that we can declare as a JavaScript function: `<field name>(<field arguments>): <field type>`.

As we wrote in our SDL example, the arguments of the `changeNickName` field can either be a list of fields. When we must deal with more and more parameters, we can use an **input type**. The `input` type object works like a `type` object, but can only be used as the user's input. It is useful when we must declare more complex input objects. Let's add to our GQL schema another mutation that accepts an `input` type:

```
input NewNickName {
  personId: ID!
  nick: String!
}
type Mutation {
  // ...
  changeNickNameWithInput(input: NewNickName!): Person!
}
```

We have defined an `input` `NewNickName`GQL Type that looks like the `Person` object type but omits the fields that the user can't set.

Well done! We have written our application GQL schema. You have seen all the basic things you need to define your GQL schemas. Before digging deeper into the GQL specification by exploring other keywords and valid syntaxes, we must consolidate our application by implementing the business logic. It is time to write some code!

How to make live a GQL schema?

In the *Writing the GQL Schema* section, we wrote the application's GQL schema. Now, we need to initialize a new npm project. For the sake of simplicity and in order to focus on the GQL logic only, we can build it by running the following code:

```
mkdir family-gql
cd family-gql/
npm init --yes
npm install fastify@4 mercurius@11
```

We are ready to create our first file, `gql-schema.js`. Here, we can just copy-paste the GQL schema we wrote in the previous section:

```
module.exports = `
# the GQL Schema string
`
```

Before proceeding further, it is worth mentioning that there are two different ways to define a GQL schema with Node.js:

- **Schema-first**: The GQL schema is a string written following the GQL specification
- **Code-first**: The GQL schema is generated by an external tool, such as the graphql npm module

In this chapter, we will follow the schema-first implementation as it is the most generic and allows you to get a clear overview of the schema without starting the application to generate it at runtime.

It is time to load the schema we wrote in the *Writing the GQL schema* section and start the GQL server. Let's see how to do it in the next section.

Starting the GQL server

To build a GQL server, we need to create a Fastify instance and register the mercurius plugin, as we have learned throughout this book. Create a new app.js file:

```
const Fastify = require('fastify')
// [1]
const mercurius = require('mercurius')
const gqlSchema = require('./gql-schema')
async function run () {
  const app = Fastify({ logger: { level: 'debug' } })
  // [2]
  const resolvers = {}
  app.register(mercurius, {
    schema: gqlSchema,
    resolvers
  })
  await app.listen({ port: 3000 })
}
run()
```

You should be able to read this small Fastify code snippet. We imported the GQL adapter and schema at line [1]. At [2], we declared an empty resolvers object and registered the mercurius plugin into the app server. If we start the application running node app.js, it will start correctly, and it will be ready to receive a GQL request.

For the sake of a test, we can run the curl command, like so:

```
curl --location --request POST 'http://localhost:3000/graphql' \
  --header 'Content-Type: application/json' \
  --data-raw '{"query":"query { family(id: 5){ id }
}","variables":{}}'
```

The command will get back an empty response:

```
{
    "data": {
        "family": null
    }
}
```

Let's step back and analyze what is happening:

1. Registering the `mercurius` plugin will expose a `/graphql` endpoint ready to receive GQL requests.

2. Since the `resolvers` option is an empty object, all the application's operations do not run any business logic, and the output is a `null` value by default.

3. Each client can execute an HTTP request to the GQL endpoint to be served. The HTTP request format is defined by the GQL specification as well.

Sending a GQL query over HTTP has very few requirements:

- There must be a `query` parameter that contains the GQL string request.

- It is possible to define a `variables` parameter that will fulfill a special placeholder in the GQL request string. It must be a plain JSON object.

- The HTTP request must be a GET or a POST method call. In the former case, the parameters must be submitted as a query-string parameter. In the latter option, the request payload must be a JSON object—as we did in our `curl` example.

If you want to know every detail about the specification, you can deep dive into this topic at the official documentation (`https://graphql.org/learn/serving-over-http/`). Mercurius supports all the specifications, so you don't need to worry about it!

> **The operationName parameter**
>
> Optionally, you can define more than one root operation type per GQL request. In this case, you need to add an extra `operationName` parameter to the HTTP request. It will select the operation to execute. An example is including a `mutation` and a `query` operation in a GQL request payload, and then specifying which one you want to run. It is often used during the development phase, and you will see it in action in the next section.

We have spun up our GQL server, but we need to add our business logic, so it is time to implement some GQL resolvers in the next section.

Implementing our first GQL query resolver

GQL resolvers implement the business logic of our application. It is possible to attach a resolver function to almost every GQL schema capability, except the root operations type. To recap, let's list all the type system's components that can have a custom resolver function:

- Type objects
- Type fields
- Scalars
- Enums
- Directives and unions

We will not discuss these topics in depth as they are out of the scope of this book.

As we discussed in the *Defining GQL operations* section, an operation is defined as a root operation type's field, so implementing the `family(id: ID!)` query will be like implementing a field's resolver.

Before continuing our journey into the resolvers implementation, we need a database to connect with. To focus on the GQL aspect of this chapter, we will apply some shortcuts to set up the fastest configuration possible to play with GQL. So, let's add an in-memory SQLite instance to our `app.js` file that will be fulfilled with mocked data at every restart. We must run the `npm install fastify-sqlite` command and then edit our application file, like so:

```
// ...
const app = Fastify({ logger: true })
await app.register(require('fastify-sqlite'), {
promiseApi: true })
await app.sqlite.migrate({
  migrationsPath: 'migrations/'
})
// ...
```

The code snippet adds the `fastify-sqlite` plugin to our Fastify application. It will connect our application to an in-memory SQLite instance by default. The module adds a convenient `migrate` utility that lets us run all the `.sql` files included in the `migrations/` directory, which we must create. In the `migrations/` folder, we can create a new `001-init.sql` file that contains our SQL schema, which recreates the tables and relations in *Figure 14.2*. Moreover, the script should add some mocked data to speed up our prototyping. You can simply copy and paste it from the book's repository at `https://github.com/PacktPublishing/Accelerating-Server-Side-Development-with-Fastify/tree/main/Chapter%2014/migrations`.

The scaffolding GQL project is ready, and now we can implement the business logic. We need to set up the `resolvers` variable we wrote in the *Starting the GQL server* section. The configuration is relatively straightforward, as follows:

```
const resolvers = {
  Query: {
    family: async function familyFunc (parent, args,
    context, info) {
      context.reply.log.info('TODO my business logic')
      return {
        id: 42,
        name: 'Demo'
      }
    }
  }
}
```

If we try to analyze the `family(id:  ID!)` query implementation, we can immediately understand the structure of the `resolvers` parameter. The `Query` key represents the **query root operation type**, and all its keys must match an entry in the corresponding root operation type defined in the GQL schema.

Matching control

During the Fastify startup, if we add a resolver function to the `Query` configuration object without declaring it in our GQL schema, then Mercurius will throw an error.

The `resolvers` object tells us that there is a query named `family`. Whenever a GQL request comes in, the GQL adapter must execute our `familyFunc` function. The resolver can be either an async or sync function that accepts four parameters:

1. `parent` is the returned value from the previous resolver. You need to know that it is always an empty object for query and mutation resolvers because there are no previous resolvers. Don't worry if it is not clear yet—we will see an example in a while.

2. `args` is a JSON parameter that contains all the query's input parameters. In our example, the family query defines a mandatory input parameter, `id`. So, `args.id` will have the user's input.

3. `context` is a shared object across all the resolvers. Mercurius builds it for every request and fulfills it with the `app` key, linking the Fastify application and the `reply` object.

4. The `info` argument represents the GraphQL node and its execution state. It is a low-level object, and you rarely need to deal with it.

Now, we can submit a query against our server, but I agree that writing a `curl` command is cumbersome. So, let's explore another cool Mercurius feature. If we tweak the configuration a bit, we can unlock a great utility:

```
app.register(mercurius, {
  schema: gqlSchema,
  graphiql: true,
  resolvers
})
```

Turning on the `graphiql` option will turn on a GraphQL client (note the *i* in `graphiql`). Running the node `app.js` application and navigating your browser to `http://127.0.0.1:3000/graphiql`, you will see the following interface:

Figure 14.3 – The graphiql user interface

So far, we wrote the GQL schema declaration, but now we need to write a GQL request string, to probe the server. The `graphiql` interface lets us play with our GraphQL server. It fetches the GQL schema and gives us autocompletion and other utilities to write GQL documents more easily.

So, we can use it to run a simple `family (id:42) { name }` query, as shown in *Figure 14.3*. The syntax aims to run the `family` query with the `id` argument set to `42`. Between the curly braces, we must list all the fields of the **object type** we want to read from the server. In this way, we are forced to select only the data we are interested in without wasting resources and keeping the payload as light as possible.

Running the query by pressing the *play* button in the center of the screen, we will trigger a call to the server, which will show the response on the right of the screen. You may notice that a `data` JSON field wraps the response payload. This structure is defined by the specification as well.

If we move to the server's logs, we should see a `TODO my business logic` log message. This means that the resolver has been executed successfully. In fact, the output shows us `name: "Demo"`, which matches the `familyFunc` function's value. Note that the query does not contain the `id` field even if the resolver returned it. This is correct, as the caller did not include it in the GQL document query.

Things are getting interesting, but it is time to run a real SQL query now:

```
const resolvers = {
  Query: {
    family: async function familyFunc (parent, args,
    context, info) {
      const sql = SQL`SELECT * FROM Family WHERE id =
        ${args.id}`
      const familyData =
        await context.app.sqlite.get(sql)
      context.reply.log.debug({ familyData }, 'Read
        familyData')
      return familyData
    }
  }
}
```

The new `familyFunc` implementation is relatively straightforward:

1. Compose the `sql` statement to execute.

2. Run the query by using the database connection through the `context.app.sqlite` decorator.

3. Log the output for debugging and return the result.

Since there is a match between the SQL `Family` table's columns and the `Family` GQL object type's fields, we can return what we get directly from the database.

Managing user input in SQL queries

You might have observed that the `sql` statement has been instantiated by the SQL tagged template utility. It is initialized like so: `const SQL = require('@nearform/sql')`. The `@nearform/sql` module provides a security layer to deal with the user input, protecting our database from SQL injection attempts.

Now, we can take a step forward by reading the family's members. To do this, we just need to modify our GQL request body:

```
query myOperationName {
  family(id: 1) {
    name
    id
    members {
      id
      fullName
    }
  }
}
```

If we run this query, we will get back an error:

```
{
  "data": {
    "family": null
  },
  "errors": [
    {
      "message": "Cannot return null for non-nullable field
       Family.members.",
      "locations": [{ "line": 5, "column": 5 }],
      "path": ["family", "members"]
    }
  ]
}
```

It is important to know that a GQL error response over HTTP always has HTTP status 200 and an `errors` array that lists all the issues found for each query you were executing. In fact, we can try to run multiple queries with a single payload:

```
query myOperationName {
  one: family(id: 1) {
    name
    id
  }
  two: family(id: 1) {
    members {
      fullName
    }
  }
}
```

The new code snippet runs two queries within the same operation name. In this case, it is mandatory to define an **alias** as we did by prepending `<label>:` before the field declaration. Note that you can use aliases to customize the response properties name of the `data` object. When writing multiple queries or mutations, you must know that queries are executed in parallel, while the server executes mutations serially instead.

The response to the previous GQL document will be similar to the first one of this section, but take a look at the `data` object:

```
{
  "data": {
    "one": {
      "name": "Foo",
      "id": "1"
    },
    "two": null
  },
  "errors": [
    {
      "message": "Cannot return null for non-nullable field
      Family.members.",
      "locations": [{"line": 9, "column": 5}],
      "path": ["two", "members"]
    }
  ]
}
```

The response payload contains the one alias with a valid output. Instead, the two property is null and has a corresponding item in the errors array.

Reading the error message, we can understand that the server is returning the null value for the Family object type, and this case conflicts with the members: [Person!]! GQL schema definition due to the exclamation mark. We defined that every Family class has at least one member, but how could we fetch them? Let's find it out in the next section.

Implementing type object resolvers

We must fetch the members of the Family object type to be able to implement the family(id: ID!) resolver correctly. We could be tempted to add another query to the familyFunc function we wrote in the *Implementing our first GQL query resolver* section. This is not the right approach to working with GQL. We must aim to build a system that is auto-consistent. This means that whatever the query operation returns, the Family type should not be aware of its relations to fulfill them, but the Family type has to know what to do. Let's edit the app.js code, as follows:

```
const resolvers = {
  Query: {
    family: async function familyFunc () { ... }
  },
  Family: {
    members: async function membersFunc (parent, args,
    context, info) {
      const sql = SQL`SELECT * FROM Person WHERE familyId
        = ${parent.id}`
      const membersData =
        await context.app.sqlite.all(sql)
      return membersData
    }
  }
}
```

The following code snippet should unveil what the GQL specification thinks. We have added a Family.members resolver by using the same pattern for the query root operation:

```
<Type object name>: {
  <field name 1>: <resolver function-1>,
  <field name n>: <resolver function-n>
}
```

So, if we read the following query, we can try to predict the resolvers that are being executed:

```
query {
  family(id: 1) {
    name
    members {
      id
      fullName
    }
  }
}
```

In this example, we can predict the following:

1. The `family` query resolver is executed, and a `Family` type is returned.

2. The `Family` type runs the `name` resolver. Since we did not define it, the default behavior is returning the property within the same name, by reading the object from *step 1*.

3. The `Family` type runs the `members` resolver. Since we have defined it, the `membersFunc` function is executed. In this case, the `parent` argument we mentioned in the *Implementing our first GQL query resolver* section equals the object returned at *step 1* of the processing. Since the `members` field expects a `Person` type array, we must return an `object` array that maps its structure.

4. The `Person` type runs the `id` and `fullName` resolvers. Since we did not define them, the default behavior is applied as described in *step 2*.

At this point, you should be astonished by the GQL power, and all the connections and possibilities it offers should be clear to you.

I will take you a little further now because we are getting a new error: `Cannot return null for non-nullable field Person.fullName.`! Let's fix it with the following `Person` type implementation:

```
const resolvers = {
  Query: { ... },
  Family: { ... },
  Person: {
    nickName: function nickNameFunc (parent) { return
    parent.nick },
    fullName: async function fullNameFunc (parent, args,
    context) {
      const sql = SQL`SELECT * FROM Family WHERE id =
        ${parent.familyId}`
      const familyData =
        await context.app.sqlite.get(sql)
```

```
        return `${parent.name} ${familyData.name}`
      },
      family: async function familyFunc (parent, args,
      context) {
        const sql = SQL`SELECT * FROM Family WHERE id =
          ${parent.familyId}`
        const familyData =
          await context.app.sqlite.get(sql)
        return familyData
      },
      friends: async function friendsFunc (parent, args,
      context) {
        const sql = SQL`SELECT * FROM Person WHERE id IN
          (SELECT friendId FROM Friend WHERE personId =
            ${parent.id})`
        const friendsData =
          await context.app.sqlite.all(sql)
        return friendsData
      }
    }
  }
```

As you can see in the previous code snippet, we have implemented all the Person fields' resolvers. The nickName function should be quite clear: the database's nick column does not match the GQL schema definition, so we need to teach the GQL adapter how to read the corresponding nickName value. The Person.family and the Person.friends resolvers apply the same logic we discussed per Family.members. Just for the sake of the experiment, the fullName resolver uses a slightly different business logic by concatenating the person's and family's names.

> **Is this design correct?**
>
> If you ask yourself if our data source is well structured, the answer is *no*. The fullName resolver is designed in a very odd way to better appreciate the optimization we will discuss in the next section.

Now, running our previous HTTP request, it should work as expected, and we can start to run complex queries like the following:

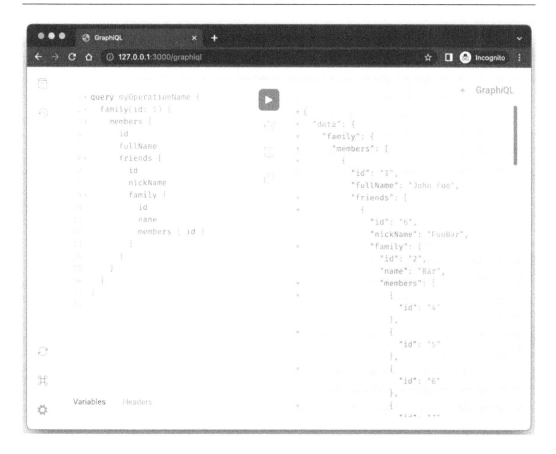

Figure 14.4 – A complex GQL query

As you can see, it is now possible to navigate the application graph exposed by our GQL server, so it is doable reading: the family members of every friend that each member of family 1 has.

Well done! Now you know how to implement any GQL resolver, you should be able to implement by yourself the mutations we defined in the *Defining GQL operations* section. The concepts are the same: you will need to add a `Mutation` property to the `resolvers` object in the `app.js` file. If you need help, you can check the solution by looking at the chapter's repository URL at `https://github.com/PacktPublishing/Accelerating-Server-Side-Development-with-Fastify/tree/main/Chapter%2014`.

You may have noticed that our implementation runs tons of SQL queries against the database for a single GQL request! This is not good at all! So, let's see how we can fix this in the next section.

How to improve resolver performance?

If we log every SQL query hitting our database when we run the GQL request in *Figure 14.4*, we will count an impressive amount of 18 queries! To do so, you can update the SQLite plugin configuration to the following:

```
const app = Fastify({ logger: { level: 'trace' } })
await app.register(require('fastify-sqlite'), {
  verbose: true,
  promiseApi: true
})
```

With the new configuration, you will be able to see all the SQL executions that have been run during the resolution of the GQL request, and you will see a lot of duplicated queries too:

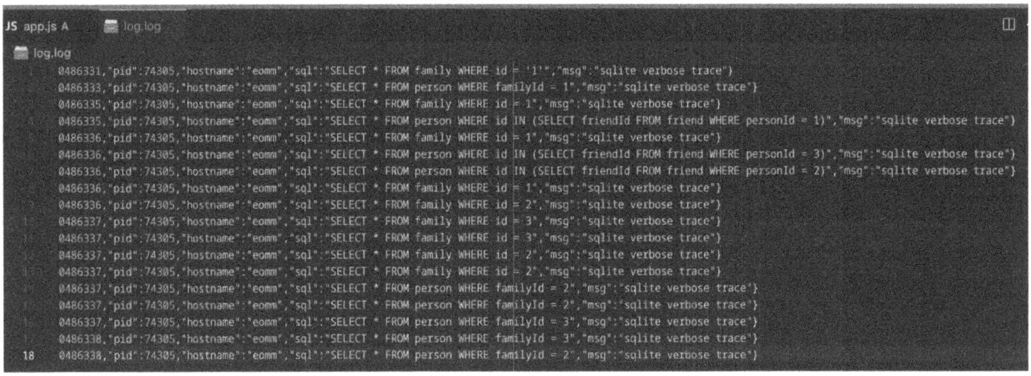

Figure 14.5 – All SQL queries run by the family resolver

This issue is called the **N+1 problem**, which ruins the service performance and wastes many server resources. For sure, a GraphQL server aims to provide simplicity over complexity. It deprecates the writing of big SQL queries with multiple joins and conditions to satisfy the relations between objects.

The solution to this everyday use case is adopting a **DataLoader**. A DataLoader is a standard data loading mechanism that manages access to the application's data sources by doing the following:

- **Batching**: This aggregates the queries executed in a single tick of the event loop or in a configured timeframe, giving you the possibility to run a single request against the data source itself
- **Caching**: This caches the queries' results per each GQL request, so if you read the same item multiple times, you will read it from the data source once

The implementation of the DataLoader API is provided by the `dataloader` npm module that you can install by running `npm install dataloader`. Let's see how to use it, after adding it to our project.

The first step is creating a new `data-loaders/family.js` file with the following code:

```
const SQL = require('@nearform/sql')
const DataLoader = require('dataloader')
module.exports = function buildDataLoader (app) {
  const dl = new DataLoader(async function fetcher (ids) {
    const secureIds = ids.map(id => SQL`${id}`)
    const sql = SQL`SELECT * FROM Family WHERE id IN
      (${SQL.glue(secureIds, ',')})`
    const familyData = await app.sqlite.all(sql)
    return ids.map((id) => familyData.find((f) => `${f.id}`
      === `${id}`))
  }, {
    cacheKeyFn: (key) => `${key}`
  })
  return dl
}
```

We built a `buildDataLoader` factory function that instantiates a new `DataLoader` instance that requires a function argument. The implementation is quite straightforward since we get an input array of `ids`, which we can use to compose a SQL query with an `IN` condition.

> **The SQL.glue method**
>
> Note that we had to write a bit more code to execute a simple SQL `IN` condition. This is necessary because SQLite does not support array parameters, as you can read in the official repository at `https://github.com/TryGhost/node-sqlite3/issues/762`. The `SQL.glue` function lets us concatenate the array of ids without losing the security checks implemented by the `@nearform/sql` module.

The most crucial logic to be aware of is the `ids.map()` phase. The `fetcher` function must return an array of results where each element matches the input IDs' order. Another important takeaway is the `cacheKeyFn` option. Since the DataLoader could be called by using string or numeric IDs, coercing all the values to strings will avoid us unpleasant unmatch due to the argument's data type—for the same reason, we are applying the same type guard when we run `familyData.find()`.

Now, you should be able to implement the `data-loaders/person.js`, `data-loaders/person-by-family.js`, and `data-loaders/friend.js` files by using the same pattern.

We have almost completed our integration. Now, we must add the DataLoaders to the GQL request's context, so we need to edit the Mercurius configuration a bit more:

```
const FamilyDataLoader = require('./data-loaders/family')
const PersonDataLoader = require('./data-loaders/person')
const PersonByFamilyDataLoader =
```

```
  require('./data-loaders/person-by-family')
const FriendDataLoader = require('./data-loaders/friend')
  // ...
  app.register(mercurius, {
    schema: gqlSchema,
    graphiql: true,
    context: async function (request, reply) {
      const familyDL = FamilyDataLoader(app)
      const personDL = PersonDataLoader(app)
      const personsByFamilyDL =
        PersonByFamilyDataLoader(app)
      const friendDL = FriendDataLoader(app)
      return { familyDL, personDL, personsByFamilyDL,
        friendDL }
    },
    resolvers
  })
```

We can extend the default context argument injected into each resolver function by setting Mercurius's context parameter.

The last step is updating the application's resolvers. Each resolver must read the data, using the DataLoaders stored in the context parameter. Here is the final result:

```
const resolvers = {
  Query: {
    family: function familyFunc (parent, args, context,
    info) {
      return context.familyDL.load(args.id)
    }
  },
  Person: {
    fullName: async function fullName (parent, args,
    context, info) {
      const familyData =
        await context.familyDL.load(parent.familyId)
      return `${parent.name} ${familyData.name}`
    },
    friends: async function friendsFunc (parent, args,
      context, info) {
      const friendsData =
        await context.friendDL.load(parent.id)
      const personsData =
      await context.personDL.loadMany(friendsData.map((f)
        => f.friendId))
      return personsData
```

```
      }
    }
  }
```

In the previous code snippet, you can see a few examples of the DataLoaders' usage. Each DataLoader offers two functions, `load` and `loadMany` to read one item or more. Now, if we read the SQLite verbose logging here, we can appreciate that the queries have been cut down by two-thirds!

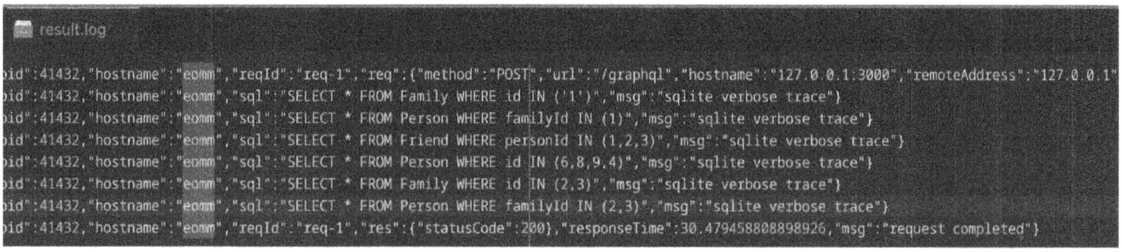

```
result.log
id":41432,"hostname":"eomm","reqId":"req-1","req":{"method":"POST","url":"/graphql","hostname":"127.0.0.1:3000","remoteAddress":"127.0.0.1"
id":41432,"hostname":"eomm","sql":"SELECT * FROM Family WHERE id IN ('1')","msg":"sqlite verbose trace"}
id":41432,"hostname":"eomm","sql":"SELECT * FROM Person WHERE familyId IN (1)","msg":"sqlite verbose trace"}
id":41432,"hostname":"eomm","sql":"SELECT * FROM Friend WHERE personId IN (1,2,3)","msg":"sqlite verbose trace"}
id":41432,"hostname":"eomm","sql":"SELECT * FROM Person WHERE id IN (6,8,9,4)","msg":"sqlite verbose trace"}
id":41432,"hostname":"eomm","sql":"SELECT * FROM Family WHERE id IN (2,3)","msg":"sqlite verbose trace"}
id":41432,"hostname":"eomm","sql":"SELECT * FROM Person WHERE familyId IN (2,3)","msg":"sqlite verbose trace"}
id":41432,"hostname":"eomm","reqId":"req-1","res":{"statusCode":200},"responseTime":30.479458808898926,"msg":"request completed"}
```

Figure 14.6 – DataLoaders aggregate the queries

The GraphQL application has been optimized! You have learned the most flexible and standard way to improve the performance of a GQL implementation. We have to mention that Mercurius supports a loader configuration option. It is less flexible than the DataLoader, but you can deep dive into this topic by reading the following blog post: `https://backend.cafe/how-to-use-dataloader-with-mercurius-graphql`.

We will now go through the last topic of this chapter to get a comprehensive overview of building a GQL server with Fastify and Mercurius: how to deal with errors?

Managing GQL errors

The GraphQL specification defines the error's format. We have seen an example of it in the *Implementing our first GQL query resolver* section. The common properties are:

- `message`: A message description

- `locations`: The GraphQL request document's coordinates that triggered the error

- `path`: The response field that encountered the error

- `extensions`: Optional field to include custom output properties

With Mercurius, we can customize the `message` error by throwing or returning an `Error` object:

```
const resolvers = {
  Query: {
    family: async function familyFunc (parent, args,
    context, info) {
      const familyData =
```

```
      await context.familyDL.load(args.id)
    if (!familyData) {
      throw new Error(`Family id ${args.id} not found`)
    }
    return familyData
  }
},
}
```

If we want to extend the errors items' field, we must follow the specification by using the extensions field. To do that with Mercurius, we must use its Error object extension. We can replace the previous code snippet with the following new one:

```
throw new mercurius.ErrorWithProps(`Family id ${args.id} not found`, {
  ERR_CODE: 404
})
```

Using the ErrorWithProps object, we can append as many properties as we need to the extensions field.

Finally, we should be aware that Fastify and Mercurius manage unhandled errors in our application to avoid memory leaks and crashes of the server. Those errors may have sensitive information, so we can add the errorFormatter option to obfuscate them:

```
app.register(mercurius, {
  errorFormatter: (result, context) => {
    result.errors = result.errors.map(hideSensitiveData)
    return mercurius.defaultErrorFormatter(result,
      context)
  },
  // [...] previous configuration
})
function hideSensitiveData (error) {
  if (error.extensions) {
    return error
  }
  error.message = 'Internal server error'
  return error
}
```

The errorFormatter function is called only when the server returns some uncaught errors. The hideSensitiveData function checks whether the error is a valid application error or a generic one. In case it is the latter, we have to overwrite the error message to preserve its locations and path metadata. Note that all the result.errors items are GraphQLError class instances, so we must differentiate the application's errors using the extensions field.

Great! We have seen all the possible error handling peculiarities to build a solid and homogeneous error response and management.

Summary

This chapter has been a dense and intensive boot camp about the GraphQL world. Now, you have a strong base theory to understand the GQL ecosystem and how it works under the hood of every GQL adapter.

We have built a GraphQL server from zero using Fastify and Mercurius. All the knowledge gained from this book's *Chapter 1* is valid because a Fastify application doesn't have any more secrets for you.

On top of your Fastify knowledge, you are able to define a GQL schema and implement its resolvers. We have solved the N+1 problem and now know the proper steps to manage it and create a fast and reliable application.

You finally have a complete toolkit at your disposal to create great APIs. You are ready to improve your code base even further by adopting TypeScript in the next chapter!

15
Type-Safe Fastify

Welcome to the final chapter of this book! This chapter will explore how Fastify's built-in TypeScript support can help us write more robust and maintainable applications.

Type safety is a crucial aspect of modern software development. It allows developers to catch errors and bugs early in the development process, reducing the time and cost of debugging and testing. By using TypeScript with Fastify, you can benefit from compile-time type safety and avoid unexpected runtime errors, leading to more stable and reliable applications.

Using TypeScript with Fastify can also improve the developer experience by providing better code completion, type inference, and documentation. In addition, Fastify has first-class support for TypeScript, providing everything needed to build robust and scalable applications, including interfaces and generics.

Moreover, using TypeScript can make deployments safer by catching potential errors and bugs before they go live. It provides an extra layer of protection to our code, giving us more confidence when deploying.

By the end of the chapter, we will have learned how to do the following:

- Create a simple Fastify project in TypeScript
- Add support for automatic type inference with so-called type-providers
- Auto-generate a documentation site for our APIs

Technical requirements

To follow along with this chapter, you will need the following:

- A working Node.js 18 installation (`https://nodejs.org/`)
- The VS Code IDE (`https://code.visualstudio.com/`)
- A Git repository is recommended but not mandatory (`https://git-scm.com/`)
- A terminal application

The code for the project can be found on GitHub at `https://github.com/PacktPublishing/ Accelerating-Server-Side-Development-with-Fastify/tree/main/ Chapter%2015`.

In the next section, we'll dive into the details of setting up the Fastify project in TypeScript, adding all of the dependencies we need to make it work.

Creating the project

Creating a new Fastify project with TypeScript support is straightforward but requires adding some extra dependencies. In this section, we'll look at the process of manually setting up a new Fastify project with TypeScript and installing the necessary dependencies.

Let's start with the `package.json` file, a configuration file for a Node.js project. It includes information about the dependencies, the entry point, and scripts. The following is just a partial snippet since we will evolve it through the sections of this chapter:

```
{
  "version": "1.0.0",
  "main": "dist/server.js", // [1]
  "dependencies": {
    "@fastify/autoload": "^5.7.1",
    "fastify": "^4.15.0"
  },
  "devDependencies": {
    "@types/node": "^18.15.11", // [2]
    "eslint-config-standard-with-typescript": "^34.0.1",
    "fastify-tsconfig": "^1.0.1", // [3]
    "rimraf": "^5.0.0",
    "tsx": "^3.12.6", // [4]
    "typescript": "^5.0.4" // [5]
  },
  "scripts": {
    "build": "rimraf dist && tsc", // [6]
    "dev": "tsx watch src/server.ts" // [7]
  }
}
```

The `package.json` file listed in the preceding code block includes the base dependencies required for the project and two scripts that will improve the development experience. As we already mentioned, to add TypeScript support, we need to add some additional development dependencies to the project besides Fastify.

Here is a breakdown of the development dependencies:

- The `main` field ([**1**]) specifies the application's entry point when it is run with the `node` . command from the project's root.

- `@types/node` ([**2**]) is a development dependency that provides TypeScript type definitions for the Node.js API. We need it to use the global variables shipped in the Node.js runtime.

- `fastify-tsconfig` ([**3**]) provides a preconfigured TypeScript configuration for use with the Fastify framework. We can extend our configuration from it and have handy defaults already configured out of the box.

- `tsx` ([**4**]) adds a TypeScript runtime tool to watch and rerun the server on file changes. It is built upon Node.js and has a zero-configuration policy.

- Finally, the `typescript` ([**5**]) development dependency adds the TypeScript compiler to check type definitions and compile the project to JavaScript. We will add a `tsconfig.json` file to the project's root to make it work properly.

Moving to the `scripts` section of `package.json`, we have the following:

- `build` ([**6**]) is a script that deletes the existing `dist` folder and invokes the TypeScript compiler (`tsc`).

- The `dev` ([**7**]) script starts the `tsx` command-line tool to rerun the application as changes are made to the project files. Running the TypeScript files directly is handy during development because it enables faster development cycles.

We are ready to create the `tsconfig.json` file in the project's root folder. This configuration file will make our Node.js project a TypeScript project.

Adding the tsconfig.json file

The `tsconfig.json` file is the configuration file for the TypeScript compiler, and it provides a way to specify options and settings that control how the code is compiled into JavaScript. For this reason, as we saw in the previous section, the Fastify team maintains the `fastify-tsconfig` npm package with the recommended configuration for Fastify plugins and applications written in TypeScript.

In the `tsconfig.json` code snippet, we can see how to use it:

```
{
  "extends": "fastify-tsconfig", // [1]
  "compilerOptions": {
    "outDir": "dist", // [2]
    "declaration": false, // [3]
    "sourceMap": true // [4]
  },
```

```
    "include": [ // [5]
      "src/**/*.ts"
    ]
}
```

Let's analyze the configuration:

- First, we use the `extends` property ([1]) to extend from `fastify-tsconfig`. This package provides a recommended configuration for Fastify web applications built with TypeScript.

- `compilerOptions` ([2]) configures the TypeScript compiler to put the compiled JavaScript files in the `dist` directory. For this reason, previously, we configured the application's entry point to `dist/server.js` using the `main` field of `package.json`.

- Since we are developing an application, our code will be run and not consumed as a library. Therefore, we set the `declaration` option to `false` ([3]) since we don't need the compiler to generate type definition files (`*.d.ts`).

- On the other hand, we want the compiler to generate source map files (`*.map`) that map the compiled JavaScript code back to the original TypeScript source code ([4]). This is useful for understanding runtime errors and debugging since it allows us to set breakpoints and step through the original TypeScript code.

- Finally, when compiling the source code, we want to include all files with the `.ts` extension inside the `src` folder and its subfolders ([5]).

Using a `tsconfig.json` file, developers can ensure that all team members use the same configuration options, providing a standardized way to configure the TypeScript compiler across different machines.

> **TypeScript compiler options**
>
> TypeScript offers a wide range of compiler options that can be specified in the `tsconfig.json` file to control the behavior of the TypeScript compiler. These options include things such as target output version, module resolution strategy, source map generation, and code generation. The Fastify team provides an opinionated configuration that is good for most projects. You can find more information about all options in the official TypeScript documentation: `https://www.typescriptlang.org/tsconfig`.

Adding the application's entry point

First, we need to write the entry point for our application. The usual Fastify server with the autoload plugin will load our routes and do the job. The code is straightforward, and we can look at it in the following snippet:

```
import { join } from 'node:path'
import Fastify from 'fastify'
```

```
import AutoLoad from '@fastify/autoload'

const fastify = Fastify({ logger: true }) // [1]

void fastify.register(AutoLoad, { // [2]
  dir: join(__dirname, 'routes')
})

fastify
  .listen({ host: '0.0.0.0', port: 3000 })
  .catch((err) => {
    fastify.log.error(err)
    process.exit(1) // [3]
  })
```

Before jumping into the code, remember to run npm i inside the project root.

Now, let's analyze the preceding snippet:

- This code creates a Fastify server with logging enabled ([1]). Since we are inside a TypeScript file, the type system is enabled. For example, if we hover over the fastify variable in the VS Code editor, we can see that it has the FastifyInstance type. Moreover, thanks to the first-class support for the TypeScript language by Fastify, everything is fully typed out of the box.

- Next, it registers a plugin using AutoLoad to load routes dynamically from the routes directory. The register method returns a Promise object, but we're not interested in its return value. By using void, we are explicitly indicating that we don't want to capture or use the return value of the Promise object, and we're just running the method for its side effects ([2]).

- Then, it starts the server on port 3000 and listens for incoming requests. If an error occurs while booting the server, it logs the error and exits the process with an error code ([3]).

Now that we've defined our entry point, we can take care of the root plugin.

Using Fastify type-providers

A Fastify type-provider is a TypeScript-only package that simplifies the definition of JSON schemas by providing type annotations and generics. Using it will allow Fastify to infer type information directly from schema definitions. Type-providers enable developers to define API endpoints' expected input and output data easily, automatically check the type correctness at compile time, and validate the data at runtime.

Fastify supports several type-providers, such as json-schema-to-ts and TypeBox. In TypeScript projects, using a type-provider can help reduce the boilerplate code required for input validation and reduce the likelihood of bugs due to invalid data types. This can ultimately make your code more robust, maintainable, and scalable.

For the sake of brevity, in the following example, we will focus only on the `TypeBox` type-provider. However, since which type-provider you use is based on personal preference, we encourage you to try other type-providers to find the best fit:

```
import { type FastifyPluginAsyncTypebox, Type } from '@fastify/type-
provider-typebox' // [1]

const plugin: FastifyPluginAsyncTypebox = async function (fastify,
_opts) { // [2]
  fastify.get('/', {
    schema: {
      querystring: Type.Object({
        name: Type.String({ default: 'world' }) // [3]
      }),
      response: {
        200: Type.Object({
          hello: Type.String() // [4]
        })
      }
    }
  }, (req) => {
    const { name } = req.query // [5]
    return { hello: name } // [6]
  })
  fastify.post('/', {
    schema: {
      body: Type.Object({
        name: Type.Optional(Type.String()) // [7]
      }),
      response: {
        200: Type.Object({
          hello: Type.String()
        })
      }
    }
  }, async (request) => {
    const { name } = request.body // [8]
    const hello = typeof name !== 'undefined' && name !==
      '' ? name : 'world'
    return { hello } // [9]
  })
}
export default plugin
```

The code snippet shows a Fastify plugin that uses `@fastify/type-provider-typebox` to define and validate the shape of the request and response objects of the routes.

Here is a breakdown of what the code does:

- First, we import `FastifyPluginAsyncTypebox` and `Type` from the `@fastify/type-provider-typebox` module ([1]). `FastifyPluginAsyncTypebox` is a type alias for `FastifyPluginAsync` that injects the support for `@sinclair/typebox` schema definitions.

- The plugin is defined as an `async` function that takes two arguments: `fastify` and `_opts`. Thanks to the explicit `FastifyPluginAsyncTypebox` type annotation ([2]), this `fastify` instance will automatically infer the types of the route schemas.

- The `fastify.get()` method defines a GET route at the root URL (`/`). We use the previously imported `Type` object to create a `querystring` object that defines a name property of the `string` type containing `"world"` as its default value ([3]). Moreover, we use it again to set the response as an object with a single `hello` property of the `string` type ([4]). Both types will automatically have the TypeScript types inferred inside the route handler.

- Hovering over the `name` variable ([5]) in VS Code will show a `string` type. This behavior happens thanks to the type-provider.

- The route handler returns an object with a single `hello` property set to the value of the `name` property extracted from the `querystring` object ([6]). The return type of the function is also inferred thanks to `TypeBox`. As an exercise, we can try changing the returning object to `{ hello: 2 }`, and the TypeScript compiler will complain since we've assigned a number instead of a string.

- The `fastify.post()` method is called to define a POST route at the root URL (`/`). The route schema includes a body object that defines an optional name property of the `string` type ([7]). Thanks to this declaration, the `request.body` object in the route handler is again fully typed ([8]). This time, we declared the `request.body.name` property as optional. We need to check whether it is `undefined` before using it in the return object and, otherwise, set it to the `world` string ([9]). As we saw for the other route handler, returning a value incompatible with the schema declaration will throw a compilation error.

Here, we wrap up this section. Thanks to type-providers, we can quickly achieve type safety across the code base by following these pointers without needing explicit type declarations:

- At runtime, the JSON schema will sanitize the inputs and serialize the outputs, making our APIs more reliable, secure, and faster.

- At compile time, our code base is fully type-checked. In addition, every variable from the schema declarations has inferred types, enabling us to find more errors before deploying the application.

In the upcoming section, we will see how we can automatically generate a documentation website compliant with the Swagger/OpenAPI specification.

Generating the API documentation

The OpenAPI specification is a widely adopted and open standard for documenting RESTful APIs. It provides a format for describing the structure and operations of an API in a machine-readable format, allowing developers to understand and interact with the API quickly.

The specification defines a set of JSON or YAML files that describe an API's endpoints, parameters, responses, and other details. This information can be used to generate API documentation, client libraries, and other tools that make it easier to work with the API.

> **Swagger and OpenAPI specifications**
>
> Swagger and OpenAPI are two related specifications, with OpenAPI being the newer version of Swagger. Swagger was originally an open source project, but later, the specification was acquired by SmartBear and renamed to OpenAPI. Today, the OpenAPI initiative, a consortium of industry leaders, maintains the specification. Swagger is also known as OpenAPI v2, while only OpenAPI generally refers to v3.

Fastify encourages developers to define JSON schemas for every route they register. It would be great if there were an automatic way to convert those definitions to the Swagger specification. And, of course, there is. But first, we must add two new dependencies to our project and use them inside the application's entry point. Now, let's install the `@fastify/swagger` and `@fastify/swagger-ui` Fastify plugins via the terminal application. To do it, in the project root, type the following command:

```
$  npm install @fastify/swagger @fastify/swagger-ui
```

Now, we can register the two newly added packages with the Fastify instance inside the `src/server.ts` file. Both packages support the Swagger and OpenAPI v3 specifications. We can choose which one to generate, passing the specific option. The following snippet configures the plugin to generate the Swagger (OpenAPI v2) specification and documentation site:

```
import { join } from 'node:path'
import Fastify from 'fastify'
import AutoLoad from '@fastify/autoload'
import Swagger from '@fastify/swagger'
import SwaggerUI from '@fastify/swagger-ui'

const fastify = Fastify({ logger: true })
void fastify.register(Swagger, {
  swagger: { // [1]
    info: { // [2]
      title: 'Hello World App Documentation',
      description: 'Testing the Fastify swagger API',
      version: '1.0.0'
    },
```

```
    consumes: ['application/json'],
    produces: ['application/json'], // [3]
    tags: [{
      name: 'Hello World', // [4]
      description: 'You can use these routes to salute
      whomever you want.'
    }]
  }
})
void fastify.register(SwaggerUI, {
  routePrefix: '/documentation' // [5]
})
// ... omitted for brevity
```

This snippet configures the `swagger` and `swagger-ui` plugins to generate the specification definitions and the documentation site. Here is a breakdown of the code:

- The `@fastify/swagger` plugin is registered. We are passing the `swagger` property to generate the specifications for OpenAPI v2 ([1]).

- We define general information about our API inside the `swagger` object, such as its title, description, and version, passing them to the `info` property ([2]). `swagger-ui` will use this to generate a website with more details.

- We define the `consumes` and `produces` arrays ([3]) to indicate the expected request and response content types. This information is crucial for the API users, and it helps when it comes to testing the endpoints.

- We define the `tags` array to group API endpoints by topic or functionality. In this case, only one tag named `Hello World` exists ([4]). In the following `src/routes/root.ts` snippet, we will see how to group the routes we have already defined.

- Finally, we register the `@fastify/swagger-ui` plugin by calling `fastify.register(SwaggerUI, {...})`. The generated documentation can then be accessed using a web browser by navigating to the URL path specified in `routePrefix` ([5]) (in this case, `/documentation`).

Next, we will modify the original route definitions to improve the auto-generated documentation. We want to do it to have better route grouping in the interface and more precise descriptions.

In the following snippet, we will omit the parts that are not relevant, but you can find the complete code in the `src/routes/root.ts` file inside the GitHub repository:

```
import { type FastifyPluginAsyncTypebox, Type } from '@fastify/type-
provider-typebox'

const plugin: FastifyPluginAsyncTypebox = async function (fastify,
```

```
_opts) {
  fastify.get('/', {
    schema: {
      tags: ['Hello World'], // [1]
      description: 'Salute someone via a GET call.', // [2]
      summary: 'GET Hello Route', // [3]
      querystring: Type.Object({
        name: Type.String({
          default: 'world',
          description: 'Pass the name of the person you
          want to salute.' // [4]
        })
      }),
  // ... omitted
  })

  fastify.post('/', {
    schema: {
      tags: ['Hello World'],
      description: 'Salute someone via a POST call.',
      summary: 'POST Hello Route',
      body: Type.Object({
        name: Type.Optional(Type.String())
      }, {
        description: 'Use the name property to pass the
        name of the person you want to salute.'
      }),
  // ... omitted
```

Even though we have shown the code additions for both routes, we will break down only the first one because the second one is structured the same:

- The tags property ([1]) specifies that the route belongs to the Hello World tag we defined while registering the @fastify/swagger plugin. This property provides a way to group related routes together in the Swagger/OpenAPI documentation.

- The description field briefly describes what the route does ([2]). It will be displayed at the top of the Swagger documentation.

- summary summarizes what the route does ([3]). It will be displayed near the route definition in the documentation.

- To provide a better understanding of the parameters accepted by an endpoint, we can add a dedicated description ([4]). It will be displayed in the Swagger documentation in the parameter details.

To test everything we added in this section, we can run our server in development mode (`npm run dev`) and use the browser to go to `http://localhost:3000/documentation`. We will be presented with a nice-looking website that we can navigate to learn more about the application we developed. Moreover, the page also integrates a client we can use to make actual calls to our API.

Summary

In this chapter, we learned why type safety is crucial in modern software development and how using TypeScript with Fastify can help catch errors and bugs early in development. The chapter also covered how using TypeScript with Fastify can improve the developer experience by providing better code completion, type inference, and the ability to auto-generate the documentation site for our APIs.

Congratulations on reaching the final chapter of this book! Throughout this journey, we have learned about the Fastify web framework and how it can help us build high-performance web applications.

Index

V

W

www.packtpub.com

Subscribe to our online digital library for full access to over 7,000 books and videos, as well as industry leading tools to help you plan your personal development and advance your career. For more information, please visit our website.

Why subscribe?

- Spend less time learning and more time coding with practical eBooks and Videos from over 4,000 industry professionals

- Improve your learning with Skill Plans built especially for you

- Get a free eBook or video every month

- Fully searchable for easy access to vital information

- Copy and paste, print, and bookmark content

Did you know that Packt offers eBook versions of every book published, with PDF and ePub files available? You can upgrade to the eBook version at www.packtpub.com and as a print book customer, you are entitled to a discount on the eBook copy. Get in touch with us at customercare@packtpub.com for more details.

At www.packtpub.com, you can also read a collection of free technical articles, sign up for a range of free newsletters, and receive exclusive discounts and offers on Packt books and eBooks.

Other Books You May Enjoy

If you enjoyed this book, you may be interested in these other books by Packt:

Supercharging Node.js Applications with Sequelize

Daniel Durante

ISBN: 978-1-80181-155-2

- Configure and optimize Sequelize for your application
- Validate your database and hydrate it with data
- Add life cycle events (or hooks) to your application for business logic
- Organize and ensure the integrity of your data even on preexisting databases
- Scaffold a database using built-in Sequelize features and tools
- Discover industry-based best practices, tips, and techniques to simplify your application development

Modern Frontend Development with Node.js

Florian Rappl

ISBN: 978-1-80461-829-5

- Develop a frontend application with Node.js-based tools and libraries
- Use bundlers such as webpack or Vite to create efficient web applications
- Leverage test runners such as Jest to ship reliable software
- Organize code repositories to work in close collaboration with other developers and teams
- Find out how to publish npm packages to simplify code reuse
- Improve code quality by enabling consistent formatting using Prettier

Packt is searching for authors like you

If you're interested in becoming an author for Packt, please visit authors.packtpub.com and apply today. We have worked with thousands of developers and tech professionals, just like you, to help them share their insight with the global tech community. You can make a general application, apply for a specific hot topic that we are recruiting an author for, or submit your own idea.

Hi!

We're Manuel, Maksim, and Matteo, the authors of Accelerating Server-Side Development with Fastify. We really hope you enjoyed reading this book and found it useful for increasing your productivity and efficiency in Fastify.

It would really help us (and other potential readers!) if you could leave a review on Amazon sharing your thoughts on Accelerating Server-Side Development with Fastify here.

Go to the link below or scan the QR code to leave your review:

https://packt.link/r/1800563582

Your review will help us to understand what's worked well in this book, and what could be improved upon for future editions, so it really is appreciated.

Best Wishes,

Manuel Spigolon

Maksim Sinik

Matteo Collina

Download a free PDF copy of this book

Thanks for purchasing this book!

Do you like to read on the go but are unable to carry your print books everywhere?

Is your eBook purchase not compatible with the device of your choice?

Don't worry, now with every Packt book you get a DRM-free PDF version of that book at no cost.

Read anywhere, any place, on any device. Search, copy, and paste code from your favorite technical books directly into your application.

The perks don't stop there, you can get exclusive access to discounts, newsletters, and great free content in your inbox daily

Follow these simple steps to get the benefits:

1. Scan the QR code or visit the link below

https://packt.link/free-ebook/9781800563582

2. Submit your proof of purchase
3. That's it! We'll send your free PDF and other benefits to your email directly

www.ingramcontent.com/pod-product-compliance
Lightning Source LLC
Chambersburg PA
CBHW062034050326
40690CB00016B/2938